T0205404

Gaming Media and Social Effects

Editor-in-chief

Henry Been-Lirn Duh, Hobart, Australia

Series editor

Anton Nijholt, Enschede, The Netherlands

More information about this series at http://www.springer.com/series/11864

Anton Nijholt

Editor

Playable Cities

The City as a Digital Playground

 Springer

Editor
Anton Nijholt
Imagineering Institute
Iskandar Puteri, Johor
Malaysia

and

Faculty of EEMCS, Human Media
    Interaction
Universiteit Twente
Enschede, Overijssel
The Netherlands

ISSN 2197-9685   ISSN 2197-9693   (electronic)
Gaming Media and Social Effects
ISBN 978-981-10-9488-0   ISBN 978-981-10-1962-3   (eBook)
DOI 10.1007/978-981-10-1962-3

Printed on acid-free paper

This Springer imprint is published by Springer Nature
The registered company is Springer Nature Singapore Pte Ltd.
The registered company address is: 152 Beach Road, #22-06/08 Gateway East, Singapore 189721, Singapore

Foreword

There are many songs of cities: how they are anthropogenic environments; how each is different from the other and produces a certain kind of mindset; how we have built them around water, foodstuffs, shelter, and less frequently but admirably, even around the idea of togetherness; how they have originated as places of trading, knowledge, and healing; how we have made them fortresses and war devices and as symbols that we can build them (and how as such, we cherish them as assets); how we have erected them around places and spaces of religious belief, including, fascinatingly, spaces of commemoration; how each and all spatialize societies and their governing rules and policies, both current, former and future; how citizens and city dwellers have challenged and overcome such policies and rules, place making; how cities, in the everyday, demonstrate an ever-changing, ever-evolving dynamic around what is public, private, harmonious, contradicting, participatory; how each city expresses notions of design, art, technology, and in the past couple of years, increasingly so, digital technology and media, and how all urbanities are, inversely, an expression of each of these; how most of us live in cities today, creating ongoing challenges for planners and citizens alike of how to handle growth, density, diversity, and so forth; and last but not least, how our cities are, and could be, playful and playable, beyond providing sufficient infrastructure, or ample green space.

Oh my, which was a built-up, labyrinthine first sentence, was not it? One that, I would hope, underlines some of the complexities that arise when addressing the research topic of Playful Cities before the possibilities brought forth by digital transformation, such as does this book contribution. Which is why it should be most useful to researchers and educators from a variety of fields, including Urban Design & Planning, Urbanism and Architecture; Human-Computer Interaction and Urban (and other types of) Computing; and, last but not least, (Serious) Game Design and other domains of behavioral design.

But what if you are now tired of reading on, and need a break? Speaking of play, why not be playful with this book instead?

We could, for example, replace a portion of the title with something silly. Or else, which leads us back to my first sentence, by replicating Dwight Garner's Paperback Game outlined in a New York Times article from 2011, we could read aloud the—hopefully—commending blurbs on this book's back cover, and let a peer come up with a first sentence that could be plausibly ascribed to this book. To live up to the book's title, we could take the edition out into your city of choice, and see if it could be used there, playfully. For example, you could hide or leave the book somewhere, for it could be the objective to find it, or to lose it, or you intend for someone unsuspecting to pick it up—perhaps a student? The book might be handy to sit on, too, for example if you're out on a plaza on a chilly, yet sunny day, observing how fellow city-dwellers play, and how your city plays back—please do not stop experimenting with how this book could be used out in the playground that is the city, I urge you!

If you are not tired of reading on, or somewhat curious, then delve into this book's many valuable approaches and examples of how cities can become playful and playable with the help of ICT, in order. e.g., to let citizens explore an urbanity; or to engage communities and to foster citizen participation; or to let citizens make, hack, be mischievous—really, to allow for us to playfully connect with our environments, and with one another, by way of our environments.

I cannot wait for you, dear reader, to smartly and responsibly build on the research presented herein. And now I'm off with my copy of this book, heading into the city, to play.

July 2016   Steffen Walz
                                                                                       RMIT University
                                                                                    Melbourne, Australia

Preface

Decades ago, the mention of 'digital cities' made us think of visually simple and relatively primitive early Internet and Worldwide Web representations of aspects of real cities, particularly in terms of culture and community. Currently, we have divergent views and expectations of digital or information and communication technology (ICT) when looking at the role that this technology can play in our ideas about our future, digitally enhanced cities. Broadband information highways have been introduced to connect countries, regions, cities, and their infrastructures worldwide. These infrastructures include governmental, business, transportation and mobility, and safety and security services. Underlying the development of these infrastructures, we have research and development on information and communication technology, research on information systems, software engineering, network technology, security and privacy, and human–computer interactions. Of course, it is possible to go into more detail and mention signal processing, nanotechnology, smart materials, machine learning, artificial intelligence, and many subareas of behavioral sciences.

In recent years, ICT has made it possible to talk about intelligent and smart cities. The focus of intelligent and smart cities can be on providing such cities with a competitive advantage over other cities. This advantage needs to come from an awareness that the comprehensive employment of advanced ICT can help to solve problems related to transportation, mobility, energy distribution, and safety and security as well as help in urban development, crucial decision-making, and stimulating innovation and economic development. Having these issues taken care of in the most efficient way is something we expect of those who develop and offer employment of smart technology.

Citizens of intelligent and smart cities may be satisfied with having an efficient home, office, and city environment. However, they usually want more, which is where issues, such as social and intimate relations, affect and emotions, playful and social interactions, entertainment, leisure and health-related needs, and playful, entertaining, recreational, and humorous activities emerge. When we discuss developments in research on intelligent and smart cities, can we also distinguish developments that address these affective, social, and playful aspects of our daily lives?

In this book, we address the issue of playfulness and playability in intelligent and smart cities. Playful technology can be introduced and authorized by city authorities. This can be compared and is similar to the introduction of smart technology in theme and recreational parks. However, smart technology becomes embedded in real-life city situations and allows real-time use by city dwellers. Thus, we can investigate how embedded smart technology can play a role in the generation and understanding of affective, playful, and humorous activities and events. Moreover, we can study how smart city technology can be used in playful, participatory design of changes in an urban environment or allow those changes to be implemented by city communities themselves.

Chapters in this book address pervasive games, urban games that change a city into a 'gameful city,' urban experiences and how to involve residents in urban city design and development. Chapters also address the playful hacking of smart city technology, mischief in smart cities, and the use of smart technology to introduce playful interactions between citizens and smart city technology in public spaces. Civic hacking that introduces playful community applications is also a topic addressed in this book. Publicly available data and infrastructure and accessible or hackable sensors and actuators can all help introduce playful applications and interactions and make an environment or city more playful and playable.

I'm grateful to the many authors who contributed to this book. They not only contributed with chapters but also acted as reviewers for two or more chapters. In addition, Dhaval Vyas and Chamari Edirisinghe helped with reviewing chapters. My thanks also go to Steffen Walz for his willingness to write a Foreword for this book. As usual, the Springer staff was very friendly and helpful, making it a pleasure to realize this publication.

Iskandar Puteri, Malaysia   Anton Nijholt
July 2016

Contents

Part III   Design for Playful Public Spaces

Towards Playful and Playable Cities

Anton Nijholt

Abstract Smart cities have inspired the introduction of various viewpoints, usually concerning the introduction of digital technology by governance bodies and its use by service providers and other economic stakeholders to allow a more efficient use of resources, transportation infrastructure, and an increase in citizen safety. However, there are also other aspects of smart cities that address, for example, our daily life activities, activities that are undertaken without having any type of efficiency in mind and interactions in which we want to engage just for social, entertainment, and fun reasons. To allow for applications that go beyond the efficient handling of our environments to those that also allow affective, social, entertaining, playful, and humorous interactions, we need to employ digital smartness. Digital smartness can be implemented in sensors and actuators in domestic situations, public spaces, and urban environments. This goal is what we want to achieve for playable cities, where residents have the opportunity to hack the city and use the smart city's data and digital technology for their own purposes and applications. Where possible, the infrastructure of a smart city can be adapted to the playful applications residents have in mind, or smart city residents can take the opportunity to hack the environment and embed their own technology in an existing global network or to introduce their own local community network. In this introduction to the Playable Cities book, various viewpoints will be discussed and short introductions to the chapters will be given.

Keywords Playable cities · Smart cities · Gameful cities · Hackable cities · Pervasive games · Sensors · Actuators

A. Nijholt (✉)
Imagineering Institute, Medini Iskandar, 79200 Iskandar Puteri, Johor, Malaysia
e-mail: anton@imagineeringinstitute.org; a.nijholt@utwente.nl

A. Nijholt
Faculty EEMCS, Human Media Interaction, University of Twente,
PO Box 217 7500 AE Enschede, The Netherlands

© Springer Science+Business Media Singapore 2017   1
A. Nijholt (ed.), *Playable Cities*, Gaming Media and Social Effects,
DOI 10.1007/978-981-10-1962-3_1

1 Introduction

The concept of playable cities was first introduced in video games such as Sim City and Grand Theft Auto (GTA) in the 1990s and the early decades of the current century. In GTA, various playable cities were implemented such as London and others that were modeled on New York City, San Francisco, Las Vegas, and Los Angeles. Playable city games such as Sim City allowed gamers to plan city development, plan, and distribute resources, regulate energy consumption, and even regulate population. These 'playable' cities rarely took into account how virtual and real residents interacted with each other or took care of their daily obligations and activities, and they usually did this without being aware or wanting to be aware of efficiency considerations.

Currently, smart technology is being introduced into our daily environment. In video game environments, we have become familiar with sensors, actuators and interaction technology (particularly in multiplayer environments). However, these sensors and actuators are embedded in a 2D or 3D graphical environment and are sometimes presented as control buttons that have to be 'pushed' or clicked, sometimes as virtual sensors that detect our proximity to a wall, an object or an opponent, or that can identify the particular direction of gaze and thereby assume interest and open up a particular opportunity or application. Clearly, we can already recognize the similar use of physical sensors and actuators embedded in our physical environments.

New technology not only makes our environment smart but also makes our clothes, skin, body, and the devices we wear smart. This technology allows us to be immersed in environments that are hardly (or not at all) related to our physical environment and appearance or to mix and augment our physical environment with a different physical or virtual environment. This technology will introduce unexpected situations with which we are not yet familiar and that confront us with surprises, not only because they are new to us but also because our conversational partners do not know how to handle them and they act in ways we do not anticipate. Maybe more importantly, this advanced information and communication technology overloads our senses, providing us with input from vision, speech and touch using different multimedia devices, including large public displays, social media displays on mobile devices, interactive pets, smart furniture, and devices in our home environments that inform us about safety, and health and energy consumption.

The concept of smart cities has been introduced to clarify how large-scale data collection of events within a city (traffic behavior, energy consumption, criminal statistics, etc.) can help to make its management more efficient by addressing efficiency issues and not just issues related to leisure, play, fun, and humor. These latter issues have been addressed in research and observations on making cities 'gameful,' hackable, playable, and playful. This book is about playful and playable cities. Its assumption is that smart technology that is available in smart and digital cities can make those cities 'playable.' That is, smart sensors and actuators that make a city smart can also be used to make a city playful and playable. By playable, we mean that not only do local authorities and city governments decide how to use and access 'big data' (data that provides information about the behavior and consumption of residents) but also how residents can introduce applications that make playful use of such information and have those applications implemented in their home or community environment to improve their daily lives. Implementing such applications can be conducted in a playful and humorous way.

In this introductory chapter, we briefly introduce the various ways designers and researchers have talked about smart cities and how they relate to playful and playable cities. In the next section (Sect. 2), we provide some views we encountered in the literature on concepts such as intelligent and smart cities. Section 3 addresses smart cities and play by focusing on the question of what playfulness can be made possible by the smartness embedded in the urban environment and surveying some opinions on playful events and activities. Section 4 focuses on pervasive games, that is, games that have been designed to be played in an urban environment. The characteristics of the environment and the way in which players moving around in it play an essential role in these games. The various aims of these games (entertainment, education, cultural exploration, tourism, urban planning) are discussed. Section 5 focuses on 'playability,' which requires environments that can be used and adapted by citizens to meet playful aims and include their ideas about playful use of their daily life. Section 6 focuses on the various projects that have made the city of Bristol (UK), an example of a playable city and includes a critical review of these projects. In Sect. 7, we reflect on how citizens can become familiar with digital technology and learn to hack their environment or take part in the design of its smart and playful adaptations.

2 Intelligent and Smart Cities

It is useful to first define what is considered intelligent, smart, playable, and playful in the context of cities. We also address humor in relation to playfulness. These notions allow us to provide views on smart digital cities that do not focus only on efficiency or consider playfulness something that contrasts with or supplements efficiency. Rather, we try to see playfulness as a characteristic of human–computer interactions that can be incorporated when designing and implementing smartness by using sensors and actuators when making this smartness available for residents and visitors of cities and city communities.

Clearly, in science fiction literature (Orwell's 'Big Brother,' Bradbury's 'Fahrenheit 451,' Huxley's 'Brave New World'), we can find views on future cities in which we have extrapolations of existing technology to applications in the home and urban environments and in which sensors and actuators help to monitor and control the activities of residents and visitors. Architects and artists provided views on future cities. Le Corbusier mentioned that "Une maison est une machine-à-habiter" (A house is a machine for living in), adding a control issue for those buildings, whether this control is by inhabitants or the 'machine.' In his New Babylon vision (Nieuwenhuys 1974) of the 1950s and 1960s, architect and artist Constant

Nieuwenhuys developed his ideas about cities as machines to play. In his 'Ludic Society,' humans are freed by automation from productive work and can follow their need to play with and transform their environments. Audio-visual media and perfected telecommunications are mentioned as important factors in this transformation to ludic social behavior. Being able to play with the elements that make up the environment is central to Constant's message. Environmental elements refer to architectural elements, elements defining the quality of space and psychological elements, that is, elements having effects on behavior and communication. Constant mentions visual, sonorous, tactile, olfactory, and gustatory aspects of these elements.

There have certainly been many others who discussed characteristics of future cities. Of course, we can quote Nicolas Tesla who, in a 1926 interview (Kennedy 1926), spoke about wireless communication that would convert the whole earth into a huge brain, or the French poet Valéry who, in 1928, predicted 'home delivery of Sensory Reality,' comparable to water and energy entering our houses through water pipes and electrical power lines (Valéry 1928). However, rather than seeing artists, architects, futurologists, and visionary scientists develop ideas about an invasion of sensory reality in our home environments or future cities, in the 1990s, we see the start of research that aims at providing a worldwide telecommunications infrastructure, including the publicly accessible 'Internet.' We see Mark Weiser's influential paper on ubiquitous computing and disappearing computers (Weiser 1991), and we also have Al Gore, serving as vice president of the United States, tell us, "We can't create a nation of information haves and have-nots. The on-ramps to the information superhighway must be accessible to all, and that will only happen if the telecommunications industry is accessible to all" (Gore 1994). In addition, at that time, there were already many initiatives that addressed the use of information and communication technology to solve problems related to the efficient use and distribution of resources in cities and cities' problems related to transportation, traffic, safety, business needs and city government and planning.

This time was also the beginning of the use of terms such as digital city, which first referred to a digital Internet or WWW representation of a real city and only later evolved into looking at a city from the viewpoint of its digital infrastructure. Other terms include 'information city,' 'invisible city' or 'intelligent city.' Singapore claimed to be transformed to an 'intelligent island' in the early 1990s (NCB 1992), and 'intelligent nations' were also introduced. Batty (1990) connected the idea of intelligent cities to a competitive advantage, a view embraced by city authorities with a focus on economic development, innovation ability and political efficiency, which could be achieved with a networked infrastructure. More elaborate views were presented in Komninos (2002) and many reports of governmental organizations, institutions, committees, and universities. More comprehensive views on intelligent cities appeared, turning intelligent cities into smart cities and adding a greater focus to urban, social, and cultural developments. Issues such as urban governance, public well-being, social cohesion, education, and (environmental) sustainability were added to the technological economic and governance viewpoints. Social, cultural, and urban development became issues. The same happened with lifestyle and leisure services.

We can find criticism on the employment of the term 'smart' and the many ways cities self-designate themselves as smart cities, especially in work by Hollands (2008). This self-designation includes attracting and catering to smart and creative workers and business. Taking Singapore as an example, it is argued that the entertainment, leisure, and lifestyle needs of these creative classes (Florida 2002) have to be serviced by a local and inexpensive workforce, thereby contributing to labor market inequalities. In Deakin and Alwaer (2011) and in Deakin (2014), more observations can be found related to the transition from intelligent to smart. This also includes more attention to the role of sensors and actuators embedded in physical environments and research concepts introduced decades earlier such as the previously mentioned disappearing computers, pervasive and ubiquitous computing and the Internet of Things. In the first decade of the twenty-first century, we see that the Internet, Worldwide Web, and social media become accessible to a larger group than a selected, privileged groups of users. We can now expect the introduction of pervasive smartness in the daily physical living environments of privileged classes, information workers and early adapters. Involving citizens in the design and development of smart technology and tuning or implementing technology to personal or community needs can be a next step. We will return to this concept in the following section.

Townsend (2014) provides a broad overview of the issues related to smart cities. He discusses how smart city technology is embraced by city departments and City Hall officials. How can we add more efficiency to a city's infrastructure? How can city departments and the companies that cooperate with them can work in more efficient ways by knowing more about the behavior of citizens and their use of energy, water, the Internet and social media, roads and public spaces? How can this knowledge, obtained by monitoring use and behavior, help to increase safety on the streets and in communities? Townsend discusses the role of the giants of information and communication technology (Cisco, IBM, Siemens) and how they team up with city officials to replace or add more technology to the existing city infrastructure. Moreover, Townsend also discusses 'bottom-up' approaches to the concept of smart cities. In these approaches, so-called civic hackers use the 'big data' that has become available from a comprehensive and integrated understanding of citizens' behavior, consumption, and activities to write programs that address problems or opportunities that are of less interest to city officials and companies but that aim to solve problems that are felt in local city communities. Hence, we can talk about 'big data' gathered and used by city departments and the companies with which they cooperate, and we can talk about citizen initiatives to use this big data to filter out information that allows them to design and implement applications that address problems felt in a particular neighborhood or social and physically distributed community and have not yet been recognized or accepted by city departments.

3   Cities that Are Smart and Playful

In addition to labels such as 'intelligent' and 'smart' as adjectives for cities, many other labels have been introduced. We are interested in the underlying descriptions that require and emphasize the role of information and communication technology. In the literature, we can find references to creative cities (Landry 2000) and 'gameful' cities (Alfrink 2014), and many authors mention playful and playable cities as well as cities with a sense of humor or humorous cities (Nijholt 2015; Edirisinghe et al. 2016). In this section, we provide some general observations on play and playfulness.

Usually, cities have designated areas for play and entertainment. Play happens in parks, particularly in public spaces where people recreate, in children's playgrounds and in places for informal sports or other outdoor activities. In Pearce (2007), it is argued that theme parks such as Disney World (Anaheim, California, 1955) were introduced as a 'narrative space,' a 'pedestrian haven for families' and 'perhaps a reaction against the suburban, freeway-interlaced sprawl that Southern California had become.' Oldenburg (2001) nevertheless attempts to recognize so-called 'Third Places' in cities. He distinguishes between First Places (our home environment), Second Places (our work environment), and Third Places, in which people gather and meet each other in a playful mood beyond the realms of home and work. Examples include coffee houses, hangouts, libraries, Internet and game cafés, and pubs, public spaces in which you connect and establish bonds with others. See also (Ferreira et al., this book).

Digitally enhancing an environment, whether it is a theme park or a Third Space, can help to make that environment more playful and more attractive. Enhancing it means making use of sensors (from detecting writing on a digital blackboard to interpreting facial expressions or other nonverbal behavior when interpreting commands or gaining knowledge about a user's preferences) and actuators that provide a user with feedback, which can be real-time (displaying the result of an explicit command) or can consist of adapting the environment and how to interact with it to the capabilities and preferences of a user based on a user's history of interactions. Can we introduce concepts such as playful, playable, and humorous in Third Spaces or should we aim to make cities playful and playable rather than designing playfulness for designated virtual or physical areas authorized by city authorities? Moreover, knowing about sensors and actuators that allow a convergence of virtual, videogame, augmented, and mixed reality and the digitally enhanced real world, how playfulness and playability be integrated into design?

Whatever architects, urban city designers and their clients or promoters wish to realize, we can expect that public and other Third Spaces will be used in ways not designed and not implemented by technology supplying corporations. This is called appropriation (Dix 2007), which can bring to mind skate boarders or graffiti creators who 'misuse' public spaces. However, we can also think of gamers who, rather than exploring a videogame environment, decide to direct their attention to exploring a smart physical environment, looking for bugs and provoking the environment, and not following routines or underlying narratives (Nijholt 2015, 2016a, b). We can also look at street artists who introduce their art in public spaces. This street art does not necessarily have the approval of city authorities, let alone their permission. Street artists or urban hacktivists such as Florian Rivière or the Singapore 'sticker lady' introduced playful changes or comments on city environments and, as was the infamous case of the sticker artist, risked caning and jail according Singapore's laws. Neither can we expect that authorities facilitate or allow 'flash mobs' or other artistic or 'counter-culture' art expressions.

Despite attempts to change the focus from efficiency issues to citizen interests and participation, there is not yet a research community that focuses on playability, playfulness, and humor in human–computer interactions. Play, playfulness, playability, and humor are concepts that are difficult to define and hard to distinguish primarily because they are related in undefined ways and whatever explanation we can think of, there will be substantial overlap between those explanations and 'definitions.' 'Play' is often associated with children's play. In agreement with Huizinga (1938) or Bateson and Martin (2013), we do not think that play is exclusive to children. Play is present when grown-up people gather together, particularly in relaxed situations, and engage in playful social interaction. 'Playfulness' is not necessarily displayed in physical interactive behavior as can be recognized in children's play. Playfulness can be about interacting with playful thoughts and ways to display them to others without necessarily becoming engaged in physical play. However, we foresee that embedded smartness, whether in devices in our environments, devices we take with us while we are on the move, or devices that are woven into our clothes and body, will allow us to visualize our playful thoughts to others and make them otherwise perceptible using multimedia that includes sound, touch, smell, and taste. Smart technology can be used to translate playful thoughts into changes in the smart environment, creating playful and humorous situations. Our playful thoughts can become available as voluntarily or even involuntarily displayed interactive multimedia events to others.

In Bateson and Martin (2013), six defining features of play are mentioned. The first five have emerged in previous studies of play by biologists and psychologists. They address issues such as spontaneity and intrinsic motivation, lack of practical goal or benefit, actions or thoughts in novel combinations, repetition of thoughts and actions, and well-being. To these features, Bateson and Martin add a sixth to define 'playful play'

> Playful play … is accompanied by a particular positive mood state in which the individual is more inclined to behave (and, in the case of humans, think) in a spontaneous and flexible way.

In particular, this latter feature is considered to define 'playfulness' in playful play. Interestingly, playing with thoughts is clearly included in the defining features of play. As mentioned previously, we now have more directly accessible sensor and actuator technology than in the past, which makes it possible for our thoughts and the playful associations we make to be realized in real-time, creating playful and humorous situations. Playfulness requires smart technology. Making a smart city playful requires citizens' access to that smart technology and the knowledge of how to use it in realizing a playful event. However, such knowledge can also be implemented in smart city technology, making it possible for this technology to facilitate, suggest or make its own decisions on how to help transform a non-playful situation into a playful situation.

In these observations, we assume that a city becomes playful (or can enhance its playfulness) through its digital smartness, regardless of whether this smartness has been provided by authorities, introduced by communities, designed and implemented by civic hackers, or realized by hacktivists in the tradition of street artists. This is not necessarily the case, as any smart technology can allow playful use, but of course, smart technology can also be developed with the aim of increasing a city's playfulness. Whether a smart city is playful depends very much on how residents experience the city and how the city stimulates 'playful play.' To make a city playful and allow for 'playful play,' it does not suffice to introduce isolated playful interactive installations (Grønbæk et al. 2012). They play the role of city authority-approved amusements or digital children's playground equipment with a fixed interaction repertoire.

4 Smart Cities as Digital Playgrounds: Pervasive Games

Cities can be smart and playful. Smartness often refers not only to attempts to have more comprehensive and efficient control of a city's resources and their distribution, traffic, transportation, and safety issues but also to attempts to provide support in decision-making and citizens' involvement in city changes and urban development plans. Available smart technology can also provide residents with the means to develop smart technology applications that are useful in their (local) community. Interfaces with these applications can be playful. These applications can also be entertaining and playful. They can also include serious and entertaining games or a mix of both.

Designers of pervasive or urban games consider the city to be a playground in which rule-based games can be designed. See also (Duggan, this book). They can be inspired by particular videogames that depend on interactions with virtual environments and virtual characters. Some of these games have been translated to (digitally enhanced) physical environments. Augmented reality and mixed reality technology can be used in the design and implementation of such games. Some early examples of videogames translated to urban environments include Augmented Reality Quake (Thomas et al. 2000) and Human Pac-Man (Cheok et al. 2003). In these games, a variety of sensors, head-mounted devices, wearable computers and wireless and Bluetooth technology is used. However, due to such advanced and expensive equipment requirements, until recently, the majority of pervasive games that are usually played by a large number of players make use of simple wearable position or proximity sensors, cameras, or microphones. There are many genres of pervasive games, and not all of them require digital technology. In Montola et al. (2009), the following genres are distinguished: (1) treasure hunts, (2) assassination games, (3) pervasive live-action role-playing games, (4) alternative reality games, (5) reality games, (6) smart street sports, (7) playful public performances, and (8) urban adventure games. For a description of these games and examples we refer to Montola et al. (2009). Especially in the last three categories, we can find examples of the use of digital technology to make challenging games.

Pervasive games can be designed and played using the available and accessible infrastructure of a city. This includes not only geographical information, public transport and traffic information, street furniture characteristics (position, function) but also information about energy and water distribution, urban development plans, and real-time information about what is happening in a city (from traffic disturbances to audience participation in a public event). Pervasive games can be fun and that can be their only aim. However, a pervasive game can also be offered to residents, an individual visitor or to a group of visitors or tourists with the aim of familiarizing them with historical, cultural or present-day information about a city environment. There can also be additional aims, such as those are related to health and well-being. Can pervasive games be introduced to persuade citizens to take part in activities that address issues of health and well-being rather than task efficiency?

Pervasive games can be introduced to understand how residents or visitors experience a city. This knowledge can be used by architects and urban city developers (designers, planners) when designing changes or extensions of city environments. A proposed change in the urban environment can be explored in a simulated virtual or augmented reality representation of that environment. These serious games, with the help of augmented reality tools, can offer design alternatives to the members of a neighborhood that has to be rebuilt. Moreover, design can become open to the people living in that neighborhood.

In addition to available wearable technology, pervasive games can of course employ sensors and actuators that are already embedded in the smart city environment. When necessary, additional digital technology tuned to a particular game can also be added. There is also the possibility to hack available technology or use it in a way that was never intended by its designers and owners. Obviously, this has been conducted with CCTV cameras that are present all over our cities. A relatively innocent form of an urban game that focuses on these cameras is the CCTV Treasure Hunt urban game developed in the UK in 2009. In this game, participants have to scout a city area and photograph as many CCTV cameras as possible. Photographs then have to be uploaded to create a map that shows a city's systems of surveillance. The aim of creating awareness of being 'spied on' is further enhanced by providing gamers with a mask to protect their anonymity while detecting these cameras. There exists a more recent version of this game, called Camover, a 'reality' game 'played' by activists in Berlin in 2013 that aimed to track and trash as many CCTV cameras possible.

Urban games have rules, thus we can ask where the playfulness and the spontaneity are. First of all, we should mention that the six features defining play as mentioned in Bateson and Martin (2013) do not contradict rule-based games. However, 'out-of-the-box' spontaneity is naturally not always appreciated by team members and by the rules of the game. Rather than being rewarded, this spontaneity would lead to punishment or exclusion from the game. However, we should also mention that pervasive games are more open to spontaneous behavior than, for example, board, exercise, serious, or sports games. Usually, players are asked to meet a particular goal, and there may be certain restrictions (location, use of devices), but generally, there are no detailed instructions on how to move or behave. Moreover, in urban or pervasive games, players have to address changing conditions in their playing field. There can be traffic jams, road works, unexpected queues in a supermarket, unknowing passersby who become involved or passersby who call the police. Pervasive game players need to handle many matters that are not defined by rules. Pervasive games do not necessarily fit in Huizinga's magic circle (Huizinga 1938).

This latter viewpoint also emerges when we look at the role of nonparticipants or bystanders of a pervasive game. In Montola et al. (2009, Chap. 10), the problem of bystanders who are unaware of the game being played in their streets is discussed. Bystanders may not view the event in a ludic context and instead perceive the events as suspicious and frightening. There are examples in which bystanders call the police to report suspicious and possibly criminal activity. One can imagine that on certain campuses in the USA, assassination games are no longer allowed. In view of increasingly omnipresent terrorist threats and warnings to the public to report unusual activity in airports, stations, public traffic, and public spaces in general, it will be increasingly difficult to play pervasive games without the permission of local authorities. Play itself can be considered an unusual activity, whatever form it takes. Obviously, pervasive games and other forms of play can be organized on commemorative days and in designated locations. Thus, one may express doubts about the future of pervasive games, particularly games that require unusual behavior or interaction with others, that is, games that require activity other than just walking around and consulting a mobile phone. In some recent surveys of pervasive games (Nevelsteen 2015; Kasapakis et al. 2013), it seems that all discussed games have been developed since 2010. Clearly, all types of location-based mobile games can be developed such that interaction with other gamers, if any, is mediated through a mobile phone rather than through physical interaction with others or the environment.

In the previous section, we already mentioned that artists who use graffiti, stickers or add confusing installations to the streets often perform controversial acts in public spaces or on buildings or street furniture. Usually, their activities do not include the use or hacking of digital technology that is embedded in the city environment. There are other movements, sometimes in the tradition of the 'Situationists' movement of the 1950s, in which citizens, sometimes with a political, societal, cultural, or community agenda, are active on the streets and organize activities that disrupt traffic or other city activities and are sometimes meant to provoke city authorities. These activities are often meant to 'reclaim the streets' for playful activities. They are not always spontaneous but also not always strictly organized, and passersby can decide to join in. Street activities can also be organized as 'flash mobs,' a sudden appearance of a group of people mobilized by messages on social media. Rather than implementing a playful activity, flash mobs currently also refer to political protests or hooligan gatherings in addition to well-rehearsed musical or theatrical performances that are not announced in advance and that are conducted in unusual places. Additionally, as has been the case for pervasive games, the police and city authorities may not allow the use of public spaces for such spontaneous events when no permit has been requested or granted.

5 Views on 'Playable Cities'

Playable Cities were mentioned in videogames such as Grand Theft Auto some decades ago. In GTA, players can perform their game actions in virtual cities that resemble existing ones such as Las Vegas, Chicago, or New York. These cities have become 'playable' in these videogames. In city simulation games, such as Cities in Motion (on transportation networks) or SimCity (on city-building and urban planning), we also see cities becoming playable. Cities in Motion have renderings of cities such as Vienna, Helsinki, Berlin, and Amsterdam. Playfulness is not really an issue as much as it is about maximizing profits. For simulation purposes, parts of existing or yet to be developed spaces and buildings have been made playable. However, from the viewpoint of this chapter and book, it is more interesting to look at playfulness and playability of real rather than virtual cities. Playability is then an issue that emerges in discussions and projects on child-friendly cities. In "10 (+1) steps to a playable city,"[1] some guidelines are presented to make cities more playable for children. Interestingly, there are no references to a possible role for ICT. Neither can we find this in German 'Lebendige Stadt' projects, including a 'Playable Town' project in a small German town and views (in 2009) on making a city 'grün und bespielbar' (green and playable).

Using smart technology to introduce playful elements in a city is one of the aims of the 'Playable City' concept that was introduced some years ago in Bristol (UK). Clearly, we can find many similar ideas in the literature and in interactive art exhibitions and events that have happened and still happen in many other cities and countries. However, it is certainly worthwhile to also support our views on a playable city by looking at the 'coining' of this term. The term 'Playable City' is considered to be a people-centered counterpoint to the idea of the data-driven 'Smart City.'

> The Playable City is imagined as a city in which hospitality and openness are key, enabling residents and visitors to reconfigure and rewrite city services, places and stories. The Playable City fosters serendipity and gives permission to be playful in public.[2]

We have a few observations about this 'definition.'

First, emphasis is placed on enabling residents and visitors to reconfigure and rewrite city services. Hence, when we take this literally, they have an active role in introducing playful situations by making changes to city services, places, and stories.

---

[1]http://www.childinthecity.eu/2015/12/08/10-1-steps-to-a-playable-city/.

[2]Call for Proposals—Making the City Playable Conference—Watershed, Bristol, UK, September 10th and 11th 2014.

It should be mentioned that in later descriptions of a playable city, the emphasis is more on the introduction of playful elements in a city that, apart from being fun to play, may help experience the city in a different way than in the routine of daily life. In later paragraphs, we return in more detail to the question of whether and how this is realized in the projects that have been introduced with the aim of making a city playable.

Second, we do not think the 'counterpoint' view is fruitful. We need smart technology to allow residents and visitors to reconfigure and rewrite city services. Smart city reconfiguring and the rewriting of city services, places, and stories requires either digital access to these services or using them in a way not necessarily foreseen by their corporate designers, developers, and software engineers. Alternatively, residents and visitors can resort to hacking sensors, actuators, services, and databases. However, it is more obvious to view residents and visitors as smart and affective nodes in the Internet of Things (IoT), that is, nodes that possess digitally enhanced human intelligence, common sense, and affect. These node characteristics allow smart or playable city residents to take advantage and enjoy the possibilities of making changes to their environments or having the environments decide on changes that suit either the users or those who have some authority over the environments. This view can also be found in Escribano (2015) who describes citizens as 'living sensors'

> I don't want to create a polarization between the extreme concepts of "Smart" or "Playable," it's not black or white, but to offer the possibility to remix, the possibility for all those involved to redefine the rules of the game constantly by using the technology as a facilitator. I am speaking about a City where their inhabitants are living sensors in a system open to changes, a reprogrammable space to challenge the body and the mind, mixed spaces of meaningful interactions and knowledge (not just about big-data at the service of big-organizations), a re-appropriated city by the citizenship, a city of pleasures (not just of efficiency) in constant development.

Third, in our observation of this 'definition' and the way it has been introduced, we want to mention some earlier visions of future cities that are related to a possible role for citizens in making changes to the environments in which they live using available or hackable data or available or hackable sensor and actuator technology. We cannot expect to see such possibilities in Huxley's Brave New World (1932) or Orwell's 1984 (1948). Both views on future cities and societies do not predict having citizens in control of their environments. In 1923, Le Corbusier introduced the view of "Une maison est une machine-à-habiter," developing a view of the house as something that either can be controlled or has control. More recently (2015), a follow-up view was presented by Sicart[3] who talks about cities as 'Machines of Play.'

Clearly, a much more developed view on future playable cities was developed in the New Babylon and 'Ludic Society' visions of Constant Nieuwenhuys (Nieuwenhuys 1974). In his view, humans, freed by automation from productive work, can follow their need to play and constantly transform their environments according to their needs. Constant assumes movable assembly systems, technical

---

[3]http://www.citymetric.com/horizons/our-cities-should-be-machines-play-869.

devices for controlling climatic conditions (light, temperature, ventilation) that are accessible to everybody, and the use of audiovisual media that becomes an important factor in ludic social behavior. About the residents of such future cities, Constant says

> They wander through the sectors of New Babylon seeking new experiences, as yet unknown ambiances. Without the passivity of tourists, but fully aware of the power they have to act upon the world, to transform it, recreate it. They dispose of a whole arsenal of technical implements for doing this, thanks to which they can make the desired changes without delay.

Currently, Constant's views can be endorsed through the facilities that advanced information and communication technology have to offer. In principle, existing technology allows users to configure or reconfigure sensors and actuators and make use of sensor data and control actuator performance in their preferred ways. Or, the other way around, embedded intelligence in sensors and actuators can learn to know about desirable changes and support the user in deciding changes or decide on them autonomously. Interacting with architectural elements that lead to physical changes in an environment is possible (Bier 2015). Robotic architecture allows users to make changes to their (or other's) physical living and playing environments. With the help of smart materials (Minuto and Nijholt 2013), the physical appearance of an environment can change because of events happening in that environment. Obviously, such events may be voluntarily created by residents or visitors in these environments. Admittedly, we are talking about technology that at this moment has hardly left the research laboratories or the living laboratories created in domestic, office, or urban environments.

6 'The' Playable City

Obviously, every city wants to be called smart and every city wants to be called playful, including cities that rank low on the possibilities citizens have to express themselves in an artistic or playful way without constraints or threats from city authorities that promise caning or imprisonment. Countries that score low on the World Press Freedom Index are happy to become involved with (authorized) initiatives to make cities more playful. Hollands (2008), in the context of smart cities, talks about a self-designated and self-congratulatory tendency, and clearly, the same observation can be made for the self-designated playful and playable cities of this decade. The question arises, what does it really mean when a city designates itself as playful or playable?

The city of Bristol (UK) has designated itself to be the world's first playable city by introducing some interactive installations in their streets during a 'playable city' period. These installations are very interesting because they are truly integrated into the streets and public spaces. From a human–computer interaction or intelligent user interface research point of view, these installations are not particularly strong or advanced. However, from an engineering point of view and being able to make robust

Fig. 1 Hello lamp post

applications accessible for the general audience in street environments, they are strong. In the following paragraphs, we describe some award-winning playable city implementations. This award was introduced in 2013 and initiated by Watershed, a film culture and digital media center in Bristol. It was supported by a network of organizations that included industry, universities and others, such as the Bristol City Council. Each year, the award-winning project is chosen from a short list of proposals submitted by design studios and artist teams.

'Hello Lamp Post' (2013)[4] was the first award-winning project (Fig. 1). Because most of the street furniture in Bristol is labeled with unique identifier codes, these codes can be used to reference locations. While walking around, items of street furniture (lamp posts, manholes, post boxes) can be 'addressed' by calling a special phone number, mentioning the code and then send a text message. Limited feedback to messages is possible because some information about the location is associated with the identifier code. Moreover, some information about previous exchanges is stored and an attempt is made to use that information in an ongoing conversation.

In 'Shadowing' (2014),[5] the second award-winning project, during several weeks, streetlights were equipped with infrared cameras that captured the shadows of pedestrians (or sometimes dogs) passing beneath and echoed them back to the next passersby who then was confronted by an unusually shaped and moving shadow (Fig. 2). Pedestrians becoming aware of this unexpected feature of the streetlights started to play with previously captured shadows and attempted to introduce unusual shadows to the system. They could also watch a streetlight recall a procession from earlier visitors.

---

[4]http://www.hellolamppost.co.uk/.

[5]http://shadowing.cc/.

Fig. 2 Shadowing

Fig. 3 Urbanimals

In 'Urbanimals' (2015),[6] the third award-winning project, different virtual animals (rabbit, kangaroo, dolphin, beetle) can be projected on walls or pavement in the streets (Fig. 3). This is occurs when a passerby is detected using computer vision. The movements of the animals take into account specific properties of the location. This has become possible by first creating a virtual copy of the physical location in which the movements can been simulated. A passerby is unexpectedly confronted with a virtual creature and may engage in interactions with the creature.

As mentioned above, the strong point of these projects is that they have been realized and made to work for some time in public spaces with access for whoever wants to play without the need for any supervision. Both Hello Lamp Post and Shadowing are realized without computer vision or position sensor technology that provides information about the behavior of their human interaction partners.

In Hello Lamp Post, the system assumes that the mobile phone user is standing in front of a particular piece of street furniture. Not a bad assumption, but you can also address the lamppost or a letter box from your home, although in that case, you may miss some remarks the system makes about the environment. The natural language processing and dialog quality is low, but the playful experience can make up for that. It can be fun to have an absurd dialog as we sometimes do when talking with a chatbot. Do not expect that the next time you are there the system will recognize you. The system does not show initiative. It does not know that passersby are interested in a conversation or even that there is a passerby.

For Shadowing, similar remarks can be made. The system collects shadows, but passersby and features of their shadows are not recognized, and the characteristics of previous passersby do not or only barely play a role in displaying one of these earlier collected shadows. Nevertheless, once playful passersby learn about the system, they can try to compose strange shadows and have them added to the repertoire of the system. Once they see what happens when they pass a lamppost displaying a shadow, they may become curious and play or have their shadow play with the displayed shadow. They will not receive feedback that is related to their own behavior.

Urbanimals knows about the presence of a passerby. Detecting this presence starts the 'behavior' of the projected animals. This behavior is canned because the program is not aware of what else is happening in the environment, thus it should not be called interaction 'with the real architecture' of the city. A similar remark can be made about the 'interactions' with passersby. How can you say that the virtual creatures are eager to play with passersby if they are not or only barely aware of the behavior of the passersby?

Whether it is Hello Lamp Post, Shadowing, or Urbanimals, feedback is either chatbot-like, canned, imagined in the mind of the user, or not present at all. From a 'state-of-the-art' research point of view, the projects are hardly interesting. Obviously, this does not prevent users from having fun while 'interacting' with these systems. Fun can follow from surprise in a 'safe' environment in which surprise can follow from seeing something unfamiliar, seeing unexpected behavior or experi-

---

[6]http://urbanimals.eu/.

encing something new. After a short period of being available and a short period of exploration by interested passersby, interest declines, and after an agreed period, such systems become dormant, are removed from the city's public spaces or remain available for tourists.

Now that we have looked at these award-winning projects, we can also ask whether such projects make a city playful or playable. Neither we, the designers, the organizers, nor the city authorities that approve such activities know. Have children been involved and in what way? Have the parents of children been involved and in what way? Who were the people who interacted with Hello Lamp Post, Shadowing, or Urbanimals? Were they people from the 'creative classes' (Florida 2002) only? How did they interact and how did they appreciate their interactions? More generally, what were the aims of each project and have there been any measurements that can indicate whether these aims have been met? Rather than formulating explicit aims and trying to validate these aims, commercial 'playable cities' projects move from one 'festival' to another, keeping city politicians, art councils, and design studios happy.

As mentioned above, there are many cities that not only want to be smart but also want to be known as playable or playful cities. For that reason, they organize 'playfulness' in parts of the city and during certain periods. They turn out to be happy with the playable city award-winning installations and want to introduce them in their cities, whether in Lagos (Nigeria), Singapore, Cape Town (South Africa), Recife (Brazil), or Tokyo (Japan). These projects can be viewed as playful or artistic projects of artists and designers that are allowed to be displayed and consumed at specific locations and during specific periods.

7 Playing, Play Making, and Urban Design

Making a city playable or playful by installing and offering authorized smart entertaining technology may make citizens happier and more satisfied with their environment. It helps to provide citizens with entertaining and healthy experiences that improve their quality of life. We can also ask whether smart technology can be offered that allows citizens to design, adapt, and (real-time) control their environments using smart sensors and actuators. Two extremes can be distinguished. One extreme (which, admittedly, is not really an extreme) is a situation in which citizens can make real-time changes to their actual environment, whether it is their home, their office or a public space. We can use our smartphone, laptop or PC, remote control or simply our presence to change our environmental conditions, to provide us with wanted information, or to generate augmented and virtual reality experiences. Another extreme is a situation in which the smart environment provides us with 'canned' and non-personalized multimedia experiences. However, we can also look at how citizens themselves can play a role in designing their environments, whether it is playful, playable or 'just' efficient and adapted to their ideas of having support while inhabiting a digitally enhanced environment. Citizens can become involved in the design or redesign of their living environment, their home environment and their public spaces. Concepts such as 'playful' and 'playable' can gain a meaning that includes citizens not only as passive consumers but also as creative users and designers of the 'playfulness' that is offered to them by authority-approved results of design and artistic teams. There is also always the possibility of hacking the digital technology that is embedded in their smart environment.

Using the technology embedded in their environment and using the data that becomes available and gathered through traditional computer bookkeeping and smart sensors, city communities can decide to improve their living environment by making changes to it. These changes can be performed based on information obtained from monitoring their environment or experimenting with sensors and actuators in their environment. This activity may require a 'maker's' or hacker's mentality and some familiarity with digital technology. We may expect a growing familiarity with such technology and a growing attention of 'makers' and 'open resource' proponents to make such technology and smart technology tool kits available for anyone interested.

Apart from citizens becoming familiar with digital technology, designers of urban environments have turned to 'participatory design' by giving a role to citizens that inhabit the environment that is under discussion. This is not new, but the employment of digital technology to realize this participation is. This employment of digital technology can take the form of games, whether they take place on digitally enhanced boards that have been designed to invoke discussions and improve decision-making or introduced in urban environments and called pervasive games, inducing gamers to explore a physical environment or virtual or augmented reality display of a future design of their daily environment. Hence, efficiency, wellness, playfulness, and playability can also be looked upon from the point of view of citizen and community participation in projects that are offered and supported by university research groups or that are activated by local communities or individuals themselves. Currently, smart urban city researchers and designers are becoming aware of the profits and interesting challenges to involving citizens in their research projects. In this book, chapters two until five are devoted urban data games and games for urban exploration.

8   About This Book

In the previous sections, we addressed many of the issues that are studied in greater detail in the next chapters of this book. Apart from this introductory chapter, the book consists of three parts. Part I examines games for playful urban design. It contains chapters on the design of games that aim to engage citizens in urban city planning, using available urban data, game concepts and ways to move around in physical, augmented and virtual reality environments. Research through design, research through making, and research through playing are issues that are addressed in these chapters. In Part II of this book, we address the design of urban and pervasive games. We examine the various characteristics that allow us to identify pervasive games, we look at digital technology that helps to develop games that make use of geographical (city) and digital information that can be accessed using mobile (movable) devices. Part III of this book addresses the design and development of playful and playable public spaces and urban environments. It also addresses participatory design, that is, the design of social and playful urban environments that include the participation of local citizens and visions for the use of sensors and actuators to design community environments that are playful and humorous. In the near future, sensors and actuators can be employed to generate playful and humorous situations and events.

Acknowledgments I'm grateful to Ben Barker (Pan Studio, Fig. 1 Hello Lamp Post), Matt Rosier (Chomko & Rosier, Fig. 2 Shadowing) and Anna Grajper (LAX: Laboratory of Architectural Experiments, Fig. 3 Urbanimals) for giving me permission to use pictures from their projects.

References

Alfrink, K.: The Gameful city. In: Walz, S.P., Deterding, S. (eds.) The Gameful World, pp. 527–560. MIT Press, London, UK (2014)

Bateson, P., Martin, P.: Play, Playfulness, Creativity and Innovation. Cambridge University Press, Cambridge (2013)

Batty, M.: Intelligent cities: using information networks to gain competitive advantage. Environ. Plan. **17**(3), 247–256 (1990). doi:10.1068/b170247

Bier, H.: Digitally-driven design and architecture. In: Harks, T., Vehlken, S. (eds.) Neighborhood Technologies, pp. 97–106. Zürich, Diaphanes (2015)

Cheok, A.D., Fong, S.W., Goh, K.H., Yang, X., Liu, W., Farzbiz, F.: Human pacman: a sensing-based mobile entertainment system with ubiquitous computing and tangible interaction. In: 2nd workshop on Network and system support for games (NetGames '03), pp. 106–117. ACM, New York (2003)

Deakin, M.: Smart cities: the state-of-the-art and governance challenge. Triple Helix **1**(7), 1–16 (2014). doi:10.1186/s40604-014-0007-9

Deakin, M., Alwaer, H.: From intelligent to smart cities. Intell. Build. Int. **3**(3), 140–152 (2011). doi:10.1080/17508975.2011.586671

Dix, A.: Designing for Appropriation. In: Proceedings of the 21st British HCI Group Annual Conference on People and Computers: HCI… but not as we know it-Volume 2, pp. 27–30. British Computer Society, London (2007)

Duggan, E.: Squaring the (Magic) Circle: A Brief Definition and History of Pervasive Games. Chapter 6, this book

Edirisinghe, C., Nijholt, A., Cheok, A.D.: From Playable to Playful: The Humorous city. In: Proceedings 8th International Conference on Intelligent Technologies for Interactive Entertainment (INTETAIN 2016), Utrecht, Netherlands, Lecture Notes of the Institute for Computer Sciences, Social Informatics and Telecommunications Engineering (LNICST). Springer, Berlin (2016)

Escribano, F.: Smart Cities, Citizen Participation & Gamification. A Starcraftian model. Online at: http://gecon.es/smart-cities-gamification/ (2015)

Ferreira, V., Anacleto, J., Bueno, A.: Designing ICT for Thirdplaceness. Chapter 10, this book

Florida, R.: The Rise of the Creative Class. Basic, New York (2002)

Gore, A.: No more information haves and have-nots. Billboard, vol. 106, issue 43, pp. 6 (1994)

Grønbæk, K., Kortbek, K.J., Møller, C., Nielsen, J., Stenfeldt, L.: Designing Playful Interactive Installations for Urban Environments—The SwingScape Experience. In: Nijholt A, Romão T, Reidsma D. (eds.) Proceedings 9th International Conference on Advances in Computer Entertainment (ACE 2012), Kathmandu, Nepal, Lecture Notes in Computer Science 7624, pp. 230–245. Springer, Heidelberg (2012)

Hollands, R.: Will the real smart city stand up? City **12**(3), 302–320 (2008)

Huizinga, J.: Homo Ludens. A Study of the Play-element in Culture. Routledge & Kegan Paul Ltd, London, UK, 1949 (First published in 1944). Also: Homo Ludens. Proeve Eener Bepaling Van Het Spel-Element Der Cultuur. H.D. Tjeenk Willink & Zoon N.V., Haarlem (1938)

Kasapakis, V., Gavalas, D., Bubaris, N.: Pervasive Games Research: A Design Aspects Based State of the Art Report. In: Proceedings of the 17th PanHellenic Conference on Informatics, pp. 152–157. ACM, New York, NY, USA (2013)

Kennedy, J.B.: When woman is boss, an interview with Nikola Tesla. Colliers (1926)

Komninos, N.: Intelligent Cities: Innovation, Knowledge Systems and Digital Spaces. Spon Press, London (2002)

Landry, C.: The Creative City: A Toolkit for Urban Innovation. Earthscan, London (2000)

Minuto, A., Nijholt, A.: Smart Material Interfaces as a Methodology for Interaction: A Survey of SMIs' State of the Art and Development. In: Proceedings of the Second International Workshop on Smart Material Interfaces: Another Step to a Material Future (SMI'13), pp. 1–6. ACM, New York, NY, USA (2013)

Montola, M., Stenros, J., Waern, A.: Pervasive Games: Theory and Design. Morgan Kaufmann Publishers Inc., San Francisco, CA, USA (2009)

NCB (National Computer Board): A Vision of an Intelligent Island: IT 2000 Report, Singapore (1992)

Nevelsteen, K.J.L.: A Survey of Characteristic Engine Features for Technology-Sustained Pervasive Games. SpringerBriefs in Computer Science, Springer, Chaim, Switzerland (2015)

Nieuwenhuys, C. (Constant): New Babylon. Manuscript. Written by Constant, for the exhibition catalogue published by the Haags Gemeentemuseum, The Hague (1974). Online at http://stichtin gconstant.nl/system/files/pdf/1974%20New%20Babylon_0.pdf

Nijholt, A.: Designing Humor for Playable Cities. In: Ahram T, Karwowski W (eds.) In: Ji YG, Choi S (eds.) Proceedings of the 6th International Conference on Applied Human Factors and Ergonomics (AHFE 2015), Section Advances in Affective and Pleasurable Design. Las Vegas, USA, Procedia Manufacturing, Volume/issue 3C, pp. 2178–2185. Elsevier (ScienceDirect) (2015)

Nijholt, A.: Mischief humor: From Games to Playable Cities. In: Proceedings 12th International Conference on Advances in Computer Entertainment Technology (ACE 2015), ACM Digital Library, Iskandar, Malaysia (2016a)

Nijholt, A.: Mischief Humor in Smart and Playable Cities. Chapter 11, this volume (2016b)

Oldenburg, R.: Celebrating the Third Place: Inspiring Stories about the "Great Good Places" at the Heart of Our Communities. Marlowe & Company, New York (2001)

Pearce, C.: Narrative environments. From Disneyland to world of warcraft. In: Von Borries, F., Walz, S.P., Böttger, M. (eds.) Space Time Play. Computer Games, Architecture and Urbanism: The Next Level. Basel, Birkhäuser (2007)

Thomas, B., Close, B., Donoghue, J., Squires, J., De Bondi, P., Morris, M., Piekarski, W.: ARQuake: an outdoor/indoor augmented reality first person application. In: 4th International Symposium on Wearable Computers, pp. 139–146. IEEE Press, New York (2000)

Townsend, A.M.: Smart Cities. W.W. Norton & Company, New York/London (2014)

Valéry, P.: The Conquest of Ubiquity. First published as "La conquête de l'ubiquité." in De La Musique avant toute chose. Editions du Tambourinaire (1928). English version in Aesthetics. Translated by Ralph Manheim. Pantheon Books, Bollingen Series, New York (1964)

Weiser, M.: The computer for the twenty-first century. Sci. Am. **265**(3), 94–10 (1991)

Part I
Games for Playful Urban Design

Games as Strong Concepts for City-Making

Ben Schouten, Gabriele Ferri, Michiel de Lange
and Karel Millenaar

Abstract Cities are becoming increasingly complex, both in terms of their social and cultural context, and in the technological solutions that are necessary to make them function. In parallel, we are observing a growing attention toward the public dimension of design, addressing societal challenges and opportunities at an urban scale. Conceptualizing, ideating, and framing design problems at a larger scale may still prove challenging, even as cities are becoming more and more relevant for all branches of design. In this chapter, we address the use of game mechanics to produce strong concepts for better understanding complex problems in city-making and communal participation, capitalizing on the necessity to shift the attention from smart cities to smart citizens. Through several examples we will show that games and play have a special quality of social bonding, providing context and motivational aspects that can be used to improve the dynamics and solutions within city-making.

B. Schouten (✉)
Department of Industrial Design, Eindhoven University of Technology,
P.O. Box 513 5600 MB Eindhoven, The Netherlands
e-mail: b.a.m.schouten@tue.nl; b.a.m.schouten@hva.nl

B. Schouten · G. Ferri · K. Millenaar
Amsterdam University of Applied Sciences, Theo Thijssenhuis, Wibautstraat 2-4,
1091 GM Amsterdam, The Netherlands
e-mail: g.ferri@hva.nl

K. Millenaar
e-mail: k.millenaar@hva.nl

M. de Lange
Department of Media and Culture Studies, Utrecht University, Muntstraat 2-2A,
3512 EV Utrecht, The Netherlands
e-mail: m.l.delange@uu.nl

© Springer Science+Business Media Singapore 2017
A. Nijholt (ed.), *Playable Cities*, Gaming Media and Social Effects,
DOI 10.1007/978-981-10-1962-3_2

23

1 Introduction

There are a growing number of applied games and playful interventions in an urban context, as a way to involve citizens and urbanites with their urban environment in the broadest possible sense, across spatial, social, and mental levels. These playful applications range from (1) Involving urbanites in the actual planning process of the city (Tan and Portugali 2012), (2) Engaging them in collective urban issues like air pollution, vacancy (de Lange 2014), (3) Engaging them with fellow citizens as a way to create more playful interactions and build trust between strangers (e.g., Kars Alfrink's Koppelkiek; 99 tiny games in UK), (4) Creating meaningful memories via playful poetic experiences (Rieser 2012), or (5) Play as critical tool, e.g., procedural rhetoric that allow people to reflect on future of their cities, or play/games as ways to imagine possible alternatives (Flanagan 2009).

This has spurred further interest in the research and design community, pushing an agenda of "playable cities", e.g., Bristol's Watershed, "playful cities" (Schouten 2015), and a more general interest in conceiving the city as a playground and urban ludic culture (Stevens 2007). Direly needed however are more in-depth reports on how these playful or gameful interventions actually "work" across the full gamut of research, conceptual, and design considerations; prototyping and testing; evaluation; portability and scaling up.

Indeed, there seems to be a disconnection between the current state of knowledge in models and visions for public services required for sustainable urban development and the implementation of this knowledge into the planning, management, and delivery of services. The development of new ways of active participation and co-creation within the optimal use of open data can help address urban challenges and bridge the implementation gap.

This chapter is organized as follows. In the next section, we will first present a concise state of the art on the need for more citizen-centered approaches to address complex urban contexts, and for processes in city-making that include different stakeholders. We will then turn to "game-making" as a resource for design research to make urban complexity more undersdable and tractable: following Stolterman and Wiberg (2010) and Höök and Löwgren (2012), we will point at game mechanics as strong concepts. We will then present data collected during the Hackable City project (Buiksloterham, Amsterdam), showing a series of three ad hoc games developed as an integral part of the inquiry process. We complement our analysis of the games with a designer interview, and we make the case for game-making effectively contributing to a more dynamic and situated understanding of complex urban scenarios.

2 Design and the Publics

Cities are becoming increasingly complex, both in terms of their social and cultural context, and of the technological solutions that are necessary to make them function. Numerous urban areas in Europe have seen a significant change in the structure and organization of public service provision (European Commission 2015). Cities are the main drivers of change in economic development and growth, knowledge production, innovation, and overall livability, recent economic, social, and environmental trends intensify the necessity to rethink traditional urban models. Different responses (from peer-to-peer informal networks, to the "sharing economy", to the "circular economy") have been recently explored to answer urban challenges, and the development of new ways of active participation and co-creation is at the center of these new models.

These design solutions are driven by a number of trends on which we elaborate: (1) a shift from smart cities to smart citizens, a focus on the end user in relation to the enabling technology, (2) open innovation focusing on process rather the simple solution making, and (3) the necessity to scale up the initial solutions to stronger concepts that can be reused and generalized.

2.1 Smart Cities and Smart Citizens

Advances in technologies such as the Internet of Things, networked sensors, and pervasive/ubiquitous computing have made digital mediations much more common in urban environments, opening new dialogues between citizens, urban planners, and design researchers in HCI, game design, interaction design, and user experience. Since the early 2000s, the term "Smart City" has been broadly adopted as a popular label to identify and cluster technology-driven approaches to urban development and renewal. Bowerman et al. (2000) characterize smart cities through their "use of advanced, integrated materials, sensors, electronics, and networks which are interfaced with computerized systems comprised of databases, tracking, and decision-making algorithms." The discourse about smart cities has been also criticized (e.g., Hollands 2008; Townsend 2013; de Lange and de Waal 2013) for a tendency to frame urban-scale design processes as top-down interventions, often technology-pushed and industry-driven, instead of bottom-up and participatory. Urban environments are characterized by social relations and by the emergence of a variety of practices. As Teli et al. (2015) have recently exemplified, designing for such complex socio-technical contexts requires taking into account the interaction between people, technologies, and places/spaces (Foth et al. 2011; Harrison and Dourish 2006).

Indeed, focusing on "smart citizens"—the inhabitants of smart cities—may prove effective in providing a more bottom-up, human-centered approach to these design spaces. "Third wave" HCI design (Bødker 2006), "design in the wild" (Dittrich et al. 2002; Rogers 2011; Chamberlain et al. 2012), as well as participatory design

(Bødker et al. 1993; Spinuzzi 2005) productively resonate with this objective. DiSalvo et al. (2014) recently provided an effective synthesis of these positions as they argue for a public orientation in design: a more public-oriented approach in design is nonsolutionist, and aims at articulating issues and giving form to problematic situations. "From this orientation, the basis for […] judgment is not whether an issue or situation resolved through design [and rather] whether we now have a better recognition and understanding of the contours of the issue or situation" (DiSalvo et al. 2014).

2.2   A Process, Involving Stakeholders

Examples of large-scale design challenges range from widespread "sharing economy" platforms (such as Uber and AirBnB for ride-sharing and apartment-sharing), to more local initiatives. For instance, about 90[1] bottom-up platforms, local digital media platforms were operating in Amsterdam in 2016, including FarmingtheCity.net, an online tool to support the creation of a local food system, or Peerby, a mobile phone app that allows people to borrow things they need from others in their vicinity. What binds these examples together is that they are part of an underlying shift in the organization of our societies, a shift we could call "the rise of the platform society" (de Waal 2014). This platform society may empower on the one hand the citizen to organize themselves around issues, bringing about a sharing economy, a participation society or civic economy. On the other hand, it forces designers to face issues of increasing complexity in terms of stakeholder diversity, pervasive technologies, and interlocking economic causes and effects.

As public-oriented design for urban spaces, the commons, and other city-scale interventions is gaining relevance, a number of design research methodologies have been leveraged to address it. Engaging such complex contexts may call for a situated approach, conceptualizing design "not as the creation of discrete, intrinsically meaningful objects, but the cultural production of new forms of practice" (Suchman et al. 1999). Participatory design (Bødker et al. 1993; Spinuzzi 2005) ideally complements situated design by actively involving people—citizens, in this specific case—in the process of ideating and creating products or services. Although participatory processes have been framed and defined in multiple ways over the years, and Vines et al. (2013) synthesize them as follows: "people are resourceful and skillful, and researchers should establish ways for this knowledge to be shared, communicated and embodied in technology design. By cooperating and forming boundary objects we provide spaces for knowledge and skills to be shared and inspire preferable future states" (Vines et al. 2013). But when it comes to articulating those opportunity spaces proposed by Vines et al. (2013), working at an urban scale pose an additional problem. Complexity increases exponentially, and so does the number of stakeholders, their needs, aspirations, and competences—as Dalsgaard (2010) shows.

---

[1]https://goo.gl/wycec6.

Framing a city-sized design problem may become an almost intractable. An inter-disciplinary dialogue with social sciences and urban planning has provided design research with useful tools—as Shklovski and Chang (2006) have argued "we are not calling for technology designers to become urban planners and social scientists, but we do suggest that there is a wealth of research in these areas that needs to be taken into account when designing new technologies" (Shklovski and Chang 2006). For example, Angus et al. (2008) leveraged an "evaluative map technique" to collect personal stories related to specific places, Resch et al. (2014) triangulated biometric data collected through smart wristbands with GPS location sensors and emotions expressed on Twitter, and Stals et al. (2014) leverage a "walking and talking" method to probe urban experiences. These, and other similarly integrated methodologies, may be used in conjunction with more traditional (non) participant ethnographic observations plus other contextual and interview methods such as panels, collective mind-mapping, or card sorting exercises. For a small selection of possible outcomes of these methodologies in urban spaces, we may point—among others—at Bentley et al. (2012), Jetter et al. (2014), Greenbaum (2010), or Lindsay et al. (2012).

2.3   Urban-Scale Challenges

The complexity of the issues at hand (problem space and solution space) constitutes a possible hindering factor in many of these examples, as well as the large community of stakeholders involved—from city council, citizens, institutions, and organiza-tions. This hinders the scaling up of these solutions as well as the development of theories and models: many solutions are only valid in a particular context and lack generalizability. Although there clearly is already a new generation of young designers, architects, activists employing digital platforms to revitalize public spaces, we are still lacking concepts and methodologies for effectively framing design issues at an urban scale. Methods for probing complex issues and making them visible, sharable, and debatable could yield more complete design insights. More engaging participatory methods could bring designers and stakeholders together, facilitating a better understanding of urban issues.

As a contribution in this sense, in this chapter we point at the strategic use of games as lenses to understand stakeholder dynamics and as a way of rapidly exploring urban issues in cooperation with stakeholders, thus producing design knowledge that can be leveraged and that can be transferred to other domains. This use of play fits into a broader approach that understands the urban public sphere not so much as a predetermined spatial site, but as a potential event space, which has recently also been embraced by city governments as a strategy to increase the quality and functionality of public spaces, a development that the UK government for science has called "vibrant cities" (Calzada 2016).

3   Game Mechanics as Strong Concepts for Urban-Scale Challenges

Given the difficulty of framing urban-scale issues in ways that are easy to grasp and generative for design/codesign insights, we turn to game-making to identify urban problems, discuss them, and ultimately understand them better together with stakeholders. Our emphasis here is on both *making* and *playing* games, as we propose that the whole process of researching, ideating, prototyping, and testing a game about urban issues may bring more clarity to how these urban issues may actually be understood and communicated. In other words, we point at games not as solutions in themselves (Morozov 2014), but at game-making as a support for design research in urban-scale issues.

The contribution we are aiming at is methodological and conceptual: we situate a *making* practice (in our specific case, making games) as an exploratory part of a much broader urban-scale design process, and we reconceptualize games neither as readymade solutions nor as final deliverables of a research project but as "thinking by doing". Making games about complex systems (e.g., urban environments) requires exploration and reflection, which in turn allows for a better understanding of the problem space. In what follows, we present and unpack our reasoning: we first look into making, hacking, and design it yourself (DIY) as forms of inquiry, then we address game design as a reflective practice, and finally we tie everything together by framing situated game-making in Stolterman and Wiberg's (2010) and Höök and Löwgren's (2012) "intermediate knowledge".

3.1   Critical Making and Game Design as a Reflective Practice

To make more tractable and understandable the complexity of urban-scale issues, we point at critical making (Ratto 2011; Ratto and Boler 2014) for an alternative way of producing and sharing design insights for/about/with a variety of stakeholders. Here we use the term "making" in the broad sense of "producing ad hoc artifacts", similarly to the related practices of hacking and DIY. Ratto (2011), who coined the term, describes critical making as "a mode of materially productive engagement that is intended to bridge the gap between creative physical and conceptual exploration." As Grimme et al. (2014) synthesize, Ratto specifically emphasizes the process of making (as opposed to the product) and material production as integral to his understanding of what critical making could be (Ratto 2011). Therefore, he presents critical making as an activity or process that seeks to reveal some sociocultural characteristics and assumptions by means of constructing ad hoc artifacts. Elaborating upon this definition, we looked at activities and artifacts that may capture complex systems and that may be easily experienced by a broad range of participants.

Flanagan (2009) has already discussed the idea of some games supporting "critical play", that is a specific way of using playful interactions to carefully examine "social, cultural, political, or even personal themes that function as alternates to popular play spaces" (2009). Our approach draws equally from critical making and from critical play. Following Flanagan, we propose playful interactions as tools to carefully examine social issues—situated, in our case, in urban spaces—and hopefully understand them better. However, we differ from Flanagan, as she points at reflective understanding as a specific effect of gameplay whereas we also draw from Ratto's proposal of inquiry through making. In sum, we are pointing at an opportunity space where (1) designing and making a game, and (2) iteratively testing it with a variety of stakeholders contribute equally to a better understanding of complex issues, such as urban-scale ones.

3.2   Toward Game Mechanics as Strong Concepts

Extracting actionable and generalizable design knowledge that connects concrete instances to abstract theories has been already discussed in design research. In a Research through Design process (Zimmerman et al. 2007), Stolterman and Wiberg (2010) point at "conceptual constructs" as midway entities that comprise the research methods applied (that is, the process of interaction design research) as well as the prototypes (that is, the products constructed according to a certain concept). The DynaBook (Kay 1972), from the early 1970s, is an example that they use in their well-known paper. The DynaBook was a concept that inspired the design of the modern laptop computer but at the time of its creation was not technologically possible to build. The original concept was to design new interaction forms, interface solutions, technology, and software for children. The DynaBook concept foresaw a low-powered flat screen that did not exist at that time. The attempts to implement the DynaBook as a prototype led to results that were too expensive, clumsy, slow, and in most ways useless. The DynaBook concept pointed in the direction of a future development toward partly digital, book-size computers. Today, we can still see the strength of this concept in, for example, Apple's naming of their laptop computer, a MacBook.

Along the same lines, Höök and Löwgren note how "design-oriented research practices create opportunities for constructing knowledge that is more abstract than particular instances, yet does not aspire to the generality of a theory" (2012), an intermediate level that contains evaluative or generative elements such as guidelines, criticism, design patterns, annotated portfolios, and strong concepts. As an example of this, we may point at citizen-led urban sensing experiments, such as the "Making Sense" project,[2] as fitting examples to these intermediate positions, in the middle

---

[2]https://www.waag.org/nl/project/making-sense. Carried out by a consortium consisting of Waag Society (Lead partner—Netherlands), University of Dundee (UK), Peer Educators Network and Science for Change (Kosovo), Institute for Advanced Architecture of Catalonia—IAAC (Spain), and the EU Joint Research Centre (Belgium).

between specific artifacts and broader theories. In this project they use the concept of citizen science, to provide the citizen with tools to change urban policies for instance in air pollution problems: Ispex (www.ispex.nl) for instance is a simple smartphone used in combination with an add-on device to measure the amount of fine dust in the air. While they are very successful in their particular application domain, it is still a big question if these initiatives and concept will lead to general practice and theory for empowering citizens in general and in other application domains.

As our specific contribution with this chapter, we argue for adding game mechanics—the algorithmic rules governing gameplay (Hunicke et al. 2004)—to Höök and Löwgren's intermediate level of design knowledge to guide the process to develop strong concepts from instances or particular problems. Games and game design have a way of producing knowledge that is clearly different from more systematized forms of design research. Games allow emphasis on sense-making in context, for ill-defined complex problems (Lombardi 2007); games are context machines that allow users to join (online) communities, so-called affinity spaces (Gee 2005; Shaffer 2006; Schouten 2015), where they can share knowledge and skills pertaining to their interest. For instance, in the various websites devoted to the game Civilization, for example, players organize themselves around the shared goal of developing expertise in the game and the skills, habits, and understandings required to successfully grow an empire—one of the victory conditions of the game.

Here, we propose that making/playing games may be useful as it transparently connects specific, concrete artifacts and issues (e.g., how rainfall and excess water is dealt with in a specific neighborhood) to much larger theoretical arguments and problems (e.g., empowerment and ownership over specific issues). Moreover, thinking and designing interactions in urban spaces has supported broader conceptualizations that are more general than a specific instance, but have not yet (or programmatically do not aim for) the breadth of a full-fledged theory.

It is important to state, that a set of game rules may be, first of all, an operational model that simplifies and represents a portion of reality from a certain point of view. It is a story or a procedure to discuss and persuade, as Bogost (2007) frames it. In the process of making a game, those rules tentatively model the general theoretical understanding of a given issue (e.g., urban empowerment) and connect it to a concrete representation (e.g., what takes place in the game). Complementarily, during play, involved stakeholders may consider the concrete elements symbolized in the game and, through strategic trial and error, explore what the game rules allow, what they forbid, which are the winning strategies (if any), thus forming a clearer mental image of the general theory behind the game. Like strong concepts and intermediate knowledge, we understand these kinds of games to have a degree of generality and not being tied to a single, specific context, thus allowing insights to emerge and be transferred also to other domains.

In other studies, strong concepts are described along four general axes: as a response to a particular use situation; triangulation between empirical, analytical, and theoretical results; horizontal grounding focusing on similarities and differences in different application domains, connecting instances and theories; and vertical grounding informing theoretical development on a more general level by broadening the context of use and usability (Höök and Löwgren 2012). While recognizing the importance of all these strands, in this work we foreground particularly the role of game mechanics in relation to horizontal and vertical grounding. More specifically, here we explore how an experienced game designer turned to a toolbox composed by a variety of mechanics to better understand a complex, urban-scale design problem. We make the case that semi-abstract game mechanics exists in the intermediate space identified by Stolterman and Wiberg (2010) and Höök and Löwgren (2012), and can be effectively leveraged to (1) abstract and model complex issues, (2) play with variables, outcomes, causes, and effects, (3) support specific "styles" of play, thus spurring reflection and discussion with stakeholders, (4) finally, grounding the issue horizontally and vertically, thus situating it in a broader context.

4   Game Mechanics for Hackable Cities

To support our argument, we present a design critique (Löwgren 2009; Chatting et al. 2015) of game-based discussion tools/probes that were developed in 2015/16 as part of the Hackable City project, in Amsterdam. In what follows, we first provide an overview of Hackable City study, of the urban context it is situated in, and of the issues it explores. Then, we outline our methodology for data collection, analysis, and interpretation. Finally, we introduce three games (Buiksloterham Matrix, the Neighborhood, The Water must Flow), we "think through" them by teasing out their game mechanics, and we triangulate our reading with a design interview with their author.

4.1   Introducing Hackable City

The Hackable City[3] is a research by design project that explores the potential for new modes of collaborative city-making in a network society. A team of academics and designers,[4] an architecture office, and societal partners explore the opportunities as well as challenges of the rise of new media technologies for an open, democratic process of collaborative city-making, which we conceptualize as the process in which citizens, designers, and institutions give shape to physical, infrastructural, legal, and social aspects of urban life (Ahlers et al. 2016; Ampatzidou et al. 2014). The Hackable City project addresses how can citizens, design professionals, local government institutions, and others employ digital media platforms in collaborative processes of urban planning, management and social organization, to contribute to

---

[3]www.thehackablecity.nl.

[4]Including the authors of this chapter.

Fig. 1 Buiksloterham, a neighborhood of Amsterdam as a field lab for circular economy (2015)

a liveable and resilient city, with a strong social fabric? The area of Buiksloterham in Amsterdam's northern part of the city (Fig. 1) is our primary case study. This is a former industrial and production area that is slated to grow from 300 to about 10,000 inhabitants in a few years. The area was given the status of an urban lab where experiments with circularity and new forms of sustainable city-making can take place, with less restrictive rules.

As making, DIY and bottom-up participative processes emerge as central topic in design research, we observe a striking parallel between the original hackers and current city makers. Like the first hackers were computer hobbyists who wrote their own software and shared it with the world, some city makers similarly contribute innovations for the existing city. Like hackers, city makers can repurpose, circumvent, or newly kickstart a range of urban infrastructures, systems, and services using fairly simple off-the-shelf digital tools. Some examples of this new sharing economy have already been mentioned under section 2 and include collaborative measuring of air quality, cooperative working spaces, collectivizing insurances and other services, producing and sharing energy, food, and other resources. To study this hacker approach to city-making, we propose to adopt a research method that is equally grounded in making. Although the Hackable City is a complex project, here we focus specifically on the part where specific games were developed as a form of design inquiry. As we will show, we adopted the practice of game-making as an alternative way of gaining insights into the complex processes of city-making. In what follows, we will present an account of how thinking through the process of game-making enabled us to obtain new perspectives. To do so, we will point at specific game mechanics as strong concepts: generative ideas that exist across domains (city, games, society, etc.) and enable us to make complex systems more tractable.

Fig. 2 The mechanics from Carcassonne (Wrede 2000) was taken to the existing situation of Buiksloterham, where future infrastructure was discussed amongst different stakeholders (citizens, city council, architects, etc.), organizing the water ecosystem of Buiksloterham, combining recreational and environmental efforts

4.2 Study Methodology

To tease out how some game mechanics operated as strong concepts in three of the Hackable City games, we turn to a mixed-method qualitative approach. We present a design critique and analysis through procedural criticism (Bogost 2006; Chatting et al. 2015) of three playable artifacts designed as part of the Hackable Cities project (Buiksloterham Matrix, the Neighborhood, The Water Must Flow), and a close reading of design interviews with their author. Other secondary materials (development diaries, sketches, playtest reports, etc.) were also considered. After a first pass of analysis, further conclusive interviews ("respondent interview") were conducted to make sure that our understanding is coherent with the designer's own interpretation of his work. The artifact-centered part of our analysis was conducted on the "finished" games, as they were demonstrated to the participants of the Hackable City project, even though these research-oriented playable artifacts are expressly fluid and often rapidly iterated upon. We produced a qualitative analysis of the three games by teasing out the game mechanics from their systems of rules, and by applying ludological categories (Aarseth 2010; Hunicke et al. 2004; Järvinen 2007) and interpreting them through procedural criticism (Bogost 2006, 2007), a design-oriented methodology for examining the relationship between the interactive, experiential, and playful qualities of an artifact and its sociocultural context (Fig. 2).

We first approached the analyses of these playable artifacts adopting a grounded theory perspective, carefully describing their mechanics and the dynamics (Hunicke et al. 2004) they supported. This was performed on a dataset composed by collected game rules, by the props and artifacts used for playing, and by coded observations from demonstrations and playtests. Furthermore, the designer expressly mentions being inspired in his creative work by game mechanics from other existing artifacts, which were adapted to address the specific issues of the Hackable City. For this reason, we include as secondary data also the other "off-the-shelf" artifacts that provided inspiration to the designer, that we consider for a comparison with the games developed for the Hackable City.

4.3 Procedural/Ludological Description of the Games

In discussing the three games that we designed (Buiksloterham Matrix, the Neighborhood, The Water Must Flow), we first outline their general structure, and then we tease out how existing game mechanics were adopted and reappropriated as a way to make the Buiksloterham context more understandable in its complexity. In other words, we present an account of how game mechanics were used as strong concepts, as a generative tool for thinking across domains (city-making, game-making, making games as design research, etc.). As strong concepts are inherently transdisciplinary and cross-media, in our description we highlight how game mechanics are reused and reappropriated from existing artifacts (from domain 1: entertainment games, to domain 2: research games), and how they illuminate and clarify specific parts of the city-making processes (from domain 2: research games, to domain 3: city-making).

Our analysis follows the chronological order in which the games were developed as part of the Hackable City research. We present (1) an experimental game to probe and reveal existing assumptions, and conceptualizations (Buiksloterham Matrix), (2) a playful experience leveraging cooperative storytelling and map-drawing to make tangible different values and related expectations (The Neighborhood), and finally (3) game mechanics for concretely demonstrating the different values and roles which could be attached to urban-scale resources (The Water Must Flow).

Buiksloterham Matrix is a tabletop game that casts players into roles that span from homeowners, local builders, public officials, and foreign investors. Gameplay is moderated by an umpire, and takes place on a large-scale printed map of the Buiksloterham neighborhood, with small tokens to show where specific actions take place. Players are assigned resources to manage, and specific goals. The overall objective is to reach one's goal within 12 turns, each of them simulating 2 months. At every turn, players declare an action to attempt and present an argument to the umpire describing why it would succeed, for example, "I am going to coordinate with my neighbors to build a small garden to reduce rainwater runoff, and I think it will easily succeed because this is a small tight-knit community where similar initiatives worked well in the past." The quality of the claims made by players are assessed by the game mediator, also with the support of a precompiled matrix with a number of variables, and turned into judgments over the likelihood that a specific event might actually take place. After this, random numbers are generated using dices, and the actions proposed by players take place (or not) according to the probabilities assigned by the umpire. Finally, the game state is updated, and the players may begin a next round by making new claims and arguments.

Buiksloterham Matrix adopts and reappropriates game mechanics from an open framework called Matrix Game System (Engle 1988), a tool for producing referee-mediated strategic games with an emphasis not on quantitative mechanics but on qualitative and rhetorical arguments. There is, however, a significant difference between Buiksloterham Matrix and the original ruleset. The standard Matrix system uses a predetermined set of labels (for example, productivity, size of the territory, diplomatic relations, etc.) in the tables helping the umpire moderating the players' claims. Buiksloterham Matrix uses a similar solution, but with a crucial difference: the variables are much more open, and players and umpire must invent the relevant categories as the game is played. If we compare the original system with this modified version, we might point at its potential usefulness as a research tool. Indeed, this game enabled designers and researchers to tease out how stakeholders conceptualized the urban environment. Observations collected during the Hackable City playtests show a slightly modified game mechanic (the matrix does not have labeled variables) generating different dynamics (Hunicke et al. 2004), prompting open conversations about the urban environment that enabled the gathering of design insights toward the following games.

For Buiksloterham Matrix, we point at "Refereed Arguments" as an emerging strong concept that is at work both in the original Matrix system and in the modified version, and that highlights a quality of bottom-up city-making. Like players have to construct their argumentations to have an impact on the game world, real-life stakeholders engage in a communal political debate at a local level, a series of bottom-up, network-like discussions with the objective of building a consensus around proposed initiatives. The strong concept of "refereed arguments" emphasizes the dialogical, social, emergent, and sometimes messy nature of this kind of processes.

The Neighborhood is the second game in this series. It is a game based on storytelling and collaborative map-drawing, and aims at generating and discussing ideas about the future of public spaces. Play is moderated by an umpire, and makes use of markers, tokens, and a large sheet of paper for cooperative drawing. Each player develops a persona that considers desirable values (e.g., sustainability, sociality, or aesthetic qualities) and problematic neighborhood issues (from noise level, to personal appropriation of communal facilities). Then, the moderator guides different rounds in which players randomly pick "event cards" and react to them by telling a story from the perspective of their persona, and by drawing on the map the locations in which it takes place. Cards contain prompts such as "Torrential rain causes the flooding of a main street", or "An exceptionally dry summer makes it difficult to water the public gardens". Similarly to an improvisational narrative exercise, the players are not allowed to directly contradict what others say, but need to co-construct a story that is coherent from a narrative point of view. All the stories are set in the same area and, as the game and the narration progress, a map of the neighborhood is collectively constructed. As they talk, players mark the map with tokens signifying problematic situations or desirable values according to their characters' description. Tokens may also be removed or replaced when specific narrative situations are resolved.

Let us now return to a comparative approach. The Neighborhood reinterprets the key game mechanics used by the narrative role-playing game The Quiet Year (2013) by Avery Mcdaldno. Differently from the artifact we are analyzing, A Quiet Year is set in a fantastic postapocalyptic world, and tasks players with describing the everyday activities of a small community of survivors. The mechanics of collaborative narration and map-drawing remain quite similar, although the overall theme of the event cards is clearly different. The value-related tokens are, instead, a significant addition that was not present in A Quiet Year. This seemingly minor change in-game mechanics has a profound impact on its emergent dynamics. By asking players to make explicit which elements are desirable and undesirable in their public spaces, The Neighborhood teases out how different values are understood and conceptualized. Also in this case, we observe an existing game mechanic being reappropriated to make visible not only the unspoken assumptions of various stakeholders but also their interplay and how they may be collectively negotiated over time.

The Neighborhood, like The Quiet Year, leverages a strong concept that we may call "Shared Narratives". The two games prompt player to cooperatively construct a story around their community and the spaces they inhabit: this mechanic points at the importance of building a shared discourse as a step toward a common identity, and at the negotiations that are necessary to do so. It also underlines the close relationship between common narratives and physical spaces, how they support and shape each other. Like in the two games, similar processes are at work also in city-making, with the social relationships between citizens producing a common discourse around specific places, which in turn give form to how they are lived, planned, and built.

Finally, The Water Must Flow is a strategic game of resource management about using water in the commons and in private areas of a neighborhood: it tasks players with managing plots of lands in relation to rainwater, flooding, draught, and other similarly water-related issues. At the beginning of the game, each player receives a stakeholder/character to role-play, with particular needs and special abilities. The main game mechanics include managing water, creating/harvesting resources (beauty, fun, livability, produce), and developing plots of land. Water, in particular, is a key component of these game mechanics. Every turn, a casual amount of rain falls upon the land plots. Each of them produces a certain amount of resources based upon how much water it receives, but only up to a limit. When water reaches the limit for a specific plot, it either overflows to an adjacent one, or it creates damage. Players need to strategically place their plots in a way that optimizes the distribution of rainwater, maximizing irrigation, and minimizing flooding.

Like the previous examples, also The Water Must Flow builds upon specific mechanics from existing games, and reinterprets them to better understand the urban context. The tile-laying mechanics, together with aesthetics qualities about building a coherent landscape with the tiles, are borrowed from Carcassonne (Wrede 2000). It is a tabletop game in which players compete for strategically arranging tiles, which generate points and benefits, and at the same time represent a rural scenario. The mechanics that deal with resource generation and allocation are adapted from Agricola (Rosenberg 2007) and Lords of Waterdeep (Lee and Thompson 2012), whose rules look at the placements of specific cards to procedurally calculate production and upkeep costs. The Water Must Flow indeed serves a twofold purpose: on one hand it elicits a playful engagement of stakeholders, but it also constitutes a conversation piece on the double role of water in neighborhood-scale planning.

For The Water Must Flow, we point at a strong concept that we may call the "Duality of Resources". Both in the existing games (Agricola, Carcassonne, Lords of Waterdeep) and in this ad hoc version, some in-game resources may play a positive and a negative role depending on how players construct their strategy. Specifically, The Water Must Flow conceptualizes rain both as a fundamental asset and as a potentially catastrophic element. Likewise, the game teases out a similar conceptualization in city-making practices, thus illuminating another way in which citizens may understand their relation to natural elements and plan their neighborhood accordingly.

As The Water Must Flow constitutes a (temporary) conclusion of our "Research through Game Making" process, it draws upon insights gathered throughout the whole series, synthesizes them, and expresses a possible overall argument. The following topics, recorded in the design diary kept during the creative process, were brought into focus through the making and testing of The Water Must Flow, and constitute its practical takeaway. Rainwater can be a threat, but also a resource in the commons, and the game system should represent this double nature. Collaboration is often an ad hoc process, not carefully planned beforehand, both in games and in urban environments. Cooperative processes are tentative, and must allow space for failure and reflection: making mistakes, stopping and starting over happens in urban practices as well as in games.

Through our practical process of game-making and playtesting with stakeholder, we were able to reflectively identify three strong concepts, which are equally at work in existing games, in the ones we designed, and in some city-making practices observed through the Hackable City research. The strong concept of "Refereed Arguments" refers to the dialogical and collective nature of designing urban spaces together, and points at the need for collective argumentations. "Shared Narratives" exemplify the construction of common identities that is frequently observed in bottom-up participatory processes, and underlines the effect that collective storytelling has on how places are conceptualized and collectively reimagined. Finally, the "Duality of Resources" highlights the complex nature of the elements of city-making: what can bring positive effects in some cases, may also cause damages in others, and citizens and planners need to be aware of this duality. We point at these observations, although not necessarily groundbreaking in the field of urban design, to exemplify how a research process based on game-making may not only bring together a variety of stakeholders, but also yield relevant insights. To complement this artifact-centered description, we now turn to a designer interview.

4.4 Designer Interview

We now present a selection of quotes from a "respondent validation debrief" conducted with Karel Millenaar,[5] one of the authors of this chapter as well as a game designer, responsible for the Hackable City games. By integrating them with theoretical insights, we support our proposal of game-making as an effective ad hoc form of inquiry into complex contexts such as urban-scale issues. Millenaar was tasked with creating applied games dealing with bottom-up urban practices, thus facing a complex, multidimensional context with no evident solution and with incomplete information. In his own words: "*I wanted to do the best job I could do, and do justice to the subject, and I wanted a deep understanding. But this would be an arrogant statement, because now I understand that [urban-scale issues] are very difficult to grasp, and not something that one person can solve. I now have a much deeper respect for the people who work on this, their professionalism, their qualities, and they're doing something so difficult.*" Through a close examination of the designer interviews and of the secondary material, we identify a pattern emerging in Millenaar's "research through game making" work. It is an ad hoc process resonating with Ratto's Critical Making (2011) and with Stolterman and Wiberg's (2010) and Höök and Löwgren's (2012) strong concepts and intermediate knowledge. We will outline here a two-pronged design strategy that is visible in our dataset. The first design tactic consists in developing and deploying games to delimit and identify the problem space, and we observe Millenaar adopting a pragmatic approach to identify where it is more productive to intervene. Second, we observe the reuse and reappropriation of game mechanics from existing off-the-shelf designs as a way for the designer to invite specific types of play and, by doing so, to situate stakeholders in significant contexts. This way, the designer argued that he was able to make the problem space more tractable and understandable. Referring to Höök and Löwgren's (2012) intermediate level of design knowledge to the ad hoc practice of game-making, we will now present extracts from the designer interview illustrating how (1) Millenaar tapped into a "toolbox" of existing game mechanics as a leverage to better frame the urban-scale context he was facing, (2) through a deliberate use of those mechanics, the three games that Millenaar developed oriented stakeholders in a more holistic understanding of the problem space, (3) ultimately Millenaar reflected back on his design process and was able to significantly reframe the problem space.

Millenaar reflects: "*My initial understanding was very vague and I didn't see any real 'issue to solve', in a sense there was a problem and in other senses there was not.*" This indeed exemplifies how urban-scale challenges may seem intractable at first, with a vague perception of some kind of underlying issue that nevertheless

---

[5]Karel Millenaar is a practicing game designer and educator based in the Netherlands. He was involved with Hackable City throughout the whole process, and provided his expertise in making, testing, and discussing the ad hoc games produced as part of the inquiry. He did not have previous experiences of academic research in the field of urban studies and urban planning, thus making his "clean-slate" approach to this problem space qualitatively representative. The designer interview presented in this paragraph was conducted by the other authors of this chapter.

proves elusive to get a grip on. Facing the difficulty of framing the context abstractly, he turns to local stakeholders as a way into the problem space. He continues: *"I was looking at how I could use the gameplay as a sort of leverage, to tell me how [people] understood their own situation, but not through talking which is always biased, and I actually wanted to know how they understood someone else's stuff - because most of the time the people dealing with these kind of projects, it's always about the interaction, not everyone is on the same page, different goals, opinions, scales ways of doing things"*.

This led to Buiksloterham Matrix, the first of the three games, which was developed to probe and tease out how stakeholders/players conceptualize the social dynamics at work in the neighborhood. We have evidence of the designer accessing his own repertoire of existing game mechanics, and choosing one that he deems promising: *"[I wanted to] figure out their understanding of the subject matter, and the matrix game makes it very easy to create these groups that are competing for something - not necessarily symmetrically.[…] I was looking at gameplay that allows people to express their understanding, and the matrix games are perfect for that. It's very simple, mechanics are not heavy, and it's not really strategic in the strictest sense."* We start seeing horizontal and vertical grounding at work here, especially with the designer selecting game-related concepts (asymmetrical competition and negotiation) and using them as a way to "think through" urban complexity. And, indeed, grounding is a fundamental component of Höök and Löwgren's strong concepts (2012). In his words: *"Not being deterministic system was important because I wanted to take the strategy away from the board, I don't want to have players focusing on playing the game successfully, I want to invite them to express the ideas they have."*

Also in the reflection on the second game, The Neighborhood, we document Millenaar picking up mechanics from other games and adapting them to foster specific effects during gameplay. *"I stumbled upon [A Quiet Year], an alternative roleplaying game with mechanics based on drawing and telling stories. […] I wanted to figure out a way for [stakeholders] to express their own [expectations] in a meaningful manner to someone else, talking about their expectations on the place they want to build a house."* And, again, we have evidence of the designer turning to his game competence to horizontally ground his understanding and have a clearer idea of the urban process: *"I brought it back to something I could understand. I delved deeper in [this process] and I realized that not every stakeholder has a plan. It's a pretty* ad hoc *process. But how can I capture this through play and learn about who everyone is? […] The Neighborhood is an experiment about creating mini-stories, mini-scenarios about a specific real location, and based on very short bursts of creativity. And for this reason I looked at A Quiet Year for an elegant and simple way to imagine a story."*

All this led to the third design, The Water Must Flow, which in a way constitutes a temporary conclusion for Millenaar's process. He reflects: *"I'm used to figure out systems that give birth to problems, I know how can I act on systems. But in this case I began to understand that water was not the actual issue, but the problem arose from something else."* Millenaar talks about a twofold approach to conceptualizing water: "*Water is a threat and, at the same time, a resource. [I aimed at having players] develop the commons and use it to create the environment that is best suited to their own fictional character, which have very different values and aspirations – for example beauty, or comfort, or luxury, or sociality...*" And another evidence of horizontal grounding through reappropriated game mechanics: "*The trick is that water is transient, it's in the commons. So I've decided to develop a standard 'upkeep' game mechanic: you need to maintain it or it's going to be destroyed, for the good or for the bad. If enough damage is done, water will fade away, an opportunity to build something new, but also an issue if you happen to need water. It's a fluid way to look into common spaces: water management can be valuable and can be expensive.*"

In conclusion, Millenaar reflects on how his own understanding the games he was making evolved throughout the process: "*I did two things: I mulled over the context, and I looked into different tabletop game mechanics. I wanted to make a game that worked as engines to get people together and sorts of interfaces to look into complexity. I wanted to put a filter over reality to make complexity easier to cope with, and make explicit the implicit. A tool that connects, deepens understanding and also transforms, all that by doing, not so much by planning. I wanted something that was meaningful to people stepping into it: how to use different mechanics to express themselves.*"

5   Discussion and Conclusion

In the same vein of practices such as critical making (Ratto 2011), in this chapter we have pointed at game-making as an ad hoc form of inquiry that can support design researchers in addressing and untangling urban-scale issues. Paraphrasing Shklovski and Chang (2006), we are not calling for technology designers, urban planners, and social scientists to become game designers, but we suggest that "thinking through" game mechanics may be particularly useful in making complex scenarios visible, understandable, and tractable.

We have reported on part of a research project entitled Hackable City, addressing some structural similarities between the "hacking" practices of early computer enthusiasts and the some innovative bottom-up city-making activities. As we faced the complexity and heterogeneity of the urban practices we were approaching, it became clear that we needed tools to make it more tractable and understandable, not only to us but also to the stakeholders themselves. By making ad hoc games, we pursued a twofold objective: on one hand, we tapped into an experienced game designer's practice to select and isolate existing game mechanics as lenses for research, and on the other one we notice how these game mechanics tease out significant parts of city-making. In other words, we argue that the mechanics we isolated act as strong concepts.

Three game mechanics/strong concepts were isolated: Refereed Arguments, Shared Narratives, and Duality of Resources. Refereed Arguments points equally at mechanics asking players to make persuasive argumentations on why a certain in-game action would have a positive outcome (as exemplified in our Buiksloterham Matrix, and in all other Matrix games), but also at the negotiations that inevitably take place in informal groups of citizens dealing with city-making processes. As game mechanics, Shared Narratives refer to collaborative story-telling by players shaping in-game elements (an element demonstrated in The Neighborhood and The Quiet Year), and also to the social process of building local identities in urban areas. Finally, games like The Water Must Flow leverage the Duality of Resources as game mechanics to simulate how certain element may have both positive and negative effects on gameplay, referring also to similar potentialities in city-making (e.g., water may pose serious problems for a city infrastructure, but may also be a significant resource). In relationship to inter-mediate knowledge and strong concepts (Stolterman and Wiberg 2010; Höök and Löwgren 2012), we focus specifically on reappropriation and reuse of game mechanics to make sense of urban phenomena by grounding them horizontally and (in part) vertically. Höök and Löwgren describe horizontal grounding as the conceptual connection between similar ideas in different domains, and the vertical one as a link between concrete instances and theories. Through the description of three games developed for the Hackable City project, and through a reflective designer interview, we have given evidence for the use of game mechanics as horizontal connectors between domains (from entertainment games to research games, and from research games to city-making), and as vertical connectors between different levels of abstraction (from the complex, multiform practices of city-making to the simplified versions enacted through games). More generally, we argue that this integrated approach can enable designers and researchers to model, simplify and communicate complex contextual elements. Clearly, urban-scale issues are not the only situations in which a similar approach may be used. More generally, we consider experimental game-making as a promising mode of inquiry for several complex application domains, as testified for example by the relative abundance of ludic simulations about (among others) economy, interna-tional policy, and geopolitics.

For most research projects in this field, the involvement of game experts has been so far principally "solutionist" (Morozov 2014)—aiming at, in other words, commu-nicating preferred solutions to well-defined problems, commonly at the end of the project itself. Instead, we have documented and presented an exploratory process that had the objective of making complex situations more understandable and acces-sible for researchers and stakeholders alike. In the Hackable City project, games were not simple "deliverables" but an integral part of the inquiry process. As city-wide design problems may risk becoming almost intractable, designers have been called to address the cultural production of new forms of practice (Suchman et al. 1999) instead of single artifacts: we see the iterative, open-ended practice of game-making clearly pushing this research agenda forward. Indeed, there are methodo-logical implications for the use of games and play in design research, especially in the construction of strong concepts. First, we foreground how games are actionable artifacts that support thinking by doing (Schouten et al. 2010), which in turn productively resonates with research through design as a form of inquiry (Zimmerman et al. 2007). We emphasize the active nature of play, as an activity that connects data and knowledge through game mechanics and, if we want to use a slogan, we might suggest that gaming, as a verb, transforms data into knowledge. As we propose this, we refer not only to playing existing games but also (as exemplified by our observations on Hackable City) through the ad hoc practice of game-making. For this reason, we draw the design research community's attention toward co-creative practices like "game jams". Similar to hackathons, game jams are very intensive (generally 24–72 h long) creative sessions where teams ideate, co-create, prototype, and pitch experimental games. Like hackathons, they have been extensively applied in the domain of therapy, healthcare, education, and research in order to balance the problem space to the solution space, to scale between the complexity of the "big" solution and the simplification of the particular problem. Game jams have been organized as part of research events—for example, the "Game Jam: [4 Research]" at CHI (Deen et al. 2014)—and studied from the perspective of design inquiry (Goddard et al. 2014). As supported by our observations in the Hackable City project, we promote extending the use of game jam-like methodologies also in other research practices, integrating game-making with more traditional forms of design inquiry. This resonates with Shaffer's (2006) conceptualization of games as affinity machines that create situated understanding: "Videogames make it possible to participate in valued communities and develop the ways of thinking that organizes these practices" thus immersing players in a socially and culturally shared experience. We point at these "affinity spaces" as opportunities generated through game-making to collectively address and discuss research issues, and to conceptualized shared solutions. In many cases this situated understanding can be transferred to other contexts—a phenomenon called the transfer of games and described by Shaffer as "epistemic frames", mechanisms through which one can use experiences in videogames to deal with situations outside the game. In our chapter, we presented evidence of game mechanics from existing (entertainment) games transferred then in an applied context, translating the motivation and experiences into the new game to model an (apparently intractable) urban context.

We see more potential in this approach and, at the same time, there is clearly much work yet to be done. Grounding is not the only trait of strong concepts, which are described along four general axes (response to a particular use situation; triangulation between empirical, analytical, and theoretical results; vertical grounding connecting instances and theories; horizontal grounding connecting related instances). Future work in this direction will include the development of a repository of core game mechanics that can be used in several application domains, followed by further research into best practices on how to perform horizontal and vertical grounding using them.

References

Aarseth, E.: Ludology. In: The Routledge Companion to Video Game Studies, pp. 484–492. Routledge (2010)

Ahlers, D., Driscoll, P., Löfström, E., Krogstie, J., Wyckmans, A.: Understanding smart cities as social machines. In: Proceedings of the 25th International Conference Companion on World Wide Web, pp. 759–764. International World Wide Web Conferences Steering Committee, Republic and Canton of Geneva, Switzerland (2016)

Ampatzidou, C., Bouw, M., van de Klundert, F., de Lange, M., de Waal, M.: The Hackable City: A Research Manifesto and Design Toolkit. Knowledge Mile, Amsterdam (2014)

Angus, A., Papadogkonas, D., Papamarkos, G., Roussos, G., Lane, G., Martin, K., West, N.,Thelwall, S., Sujon, Z., Silverstone, R.: Urban social tapestries. IEEE Pervasive Comput. 7, 44–51 (2008)

Bentley, F., Cramer, H., Hamilton, W., Basapur, S.: Drawing the City: Differing Perceptions of the Urban Environment. In: Proceedings of the SIGCHI Conference on Human Factors in Computing Systems, pp. 1603–1606. ACM, New York, NY, USA (2012)

Bødker, S.: When Second Wave HCI Meets Third Wave Challenges. In: Proceedings of the 4th Nordic Conference on Human-computer Interaction: Changing Roles, pp. 1–8. ACM, New York, NY, USA (2006)

Bødker, S., Grønbæk, K., Kyng, M.: Cooperative Design: Techniques and Experiences from the Scandinavian Scene. In: Participatory design. principles and practices. Lawrence Erlbaum Associates, Mahwah, New Jersey (1993)

Bogost, I.: Unit Operations: An Approach to Videogame Criticism. MIT Press, Cambridge, MA (2006)

Bogost, I.: Persuasive Games: The Expressive Power of Videogames. MIT Press, Cambridge, MA (2007)

Bowerman, B., Braverman, J., Taylor, J., Todosow, H., Von Wimmersperg, U.: The Vision of a Smart City. In: 2nd International Life Extension Technology Workshop, pp. 48–58. Paris (2000)

Calzada, I.: Smartness for Prosperity: UK & Brazil. Social Science Research Network, Rochester, NY (2016)

Chamberlain, A., Crabtree, A., Rodden, T., Jones, M., Rogers, Y.: Research in the Wild: Understanding "in the Wild" Approaches to Design and Development. In: Proceedings of the Designing Interactive Systems Conference, pp. 795–796. ACM, New York, NY, USA (2012)

Chatting, D., Kirk, D.S., Yurman, P., Bichard, J.-A.: Designing for Family Phatic Communication: A Design Critique Approach. In: Proceedings of the 2015 British HCI Conference, pp. 175–183. ACM, New York, NY, USA (2015)

Dalsgaard, P.: Challenges of Participation in Large-scale Public Projects. In: Proceedings of the 11th Biennial Participatory Design Conference, pp. 21–30. ACM, New York, NY, USA (2010)

de Lange, M., de Waal, M.: Owning the city: New media and citizen engagement in urban design. First Monday. 18 (2013)

de Lange, M.: Playful Planning: Citizens Making The Smart And Social City. ECLECTIS report: A contribution from cultural and creative actors to citizens' empowerment (2014)

Deen, M., Cercos, R., Chatman, A., Naseem, A., Bernhaupt, R., Fowler, A., Schouten, B., Mueller, F.: Game Jam: [4 Research]. In: CHI '14 Extended Abstracts on Human Factors in Computing Systems, pp. 25–28. ACM, New York, NY, USA (2014)

de Waal, M.: The City as Interface: How Digital Media are Changing The City. Nai010 Publishers, Rotterdam, Netherlands (2014)

DiSalvo, C., Lukens, J., Lodato, T., Jenkins, T., Kim, T.: Making Public Things: How HCI Design Can Express Matters of Concern. In: Proceedings of the 32nd Annual ACM Conference on Human Factors in Computing Systems, pp. 2397–2406. ACM, New York, NY, USA (2014)

Dittrich, Y., Eriksén, S., Hansson, C.: PD in the Wild. Evolving Practices of Design in Use. Presented at the Participatory Design Conference (2002)

Engle, C.: Matrix Games Systems (game). Chris Engle, Kansas City (1988)

European Commission.: European Research Area: Global Urban Challenges, Joint European Solutions. Cofund Smart Urban Futures Call Text. http://jpi-urbaneurope.eu (2015)

Flanagan, M.: Critical Play: Radical Game Design. MIT Press, Cambridge MA (2009)

Foth, M.: From Social Butterfly To Engaged Citizen: Urban Informatics, Social Media, Ubiquitous Computing, and Mobile Technology to Support citizen Engagement. MIT Press, Cambridge MA (2011)

Gee, J.P.: Semiotic social spaces and affinity spaces: from the age of mythology to today's schools. In: Beyond Communities of Practice. Cambridge University Press, Cambridge (2005)

Goddard, W., Byrne, R., Mueller, F.: Playful Game Jams: Guidelines for Designed Outcomes. In: Proceedings of the 2014 Conference on Interactive Entertainment. p. 6:1–6:10. ACM, New York, NY, USA (2014)

Greenbaum, J.: Situations and interactions: digital CafÉ squatting and participatory design. In: Proceedings of the 11th Biennial Participatory Design Conference, pp. 243–246. ACM, New York, NY, USA (2010)

Grimme, S., Bardzell, J., Bardzell, S.: "We've Conquered Dark": Shedding light on empowerment in critical making. In: Proceedings of the 8th Nordic Conference on Human-Computer Interaction: Fun, Fast, Foundational, pp. 431–440. ACM, New York, NY, USA (2014)

Harrison, S., Dourish, P.: Re-place-ing Space: The Roles of Place and Space in Collaborative Systems. In: Proceedings of the 1996 ACM Conference on Computer Supported Cooperative Work, pp. 67–76. ACM, New York, NY, USA (1996)

Hollands, R.G.: Will the real smart city please stand up? City 12, 303–320 (2008)

Höök, K., Löwgren, J.: Strong Concepts: Intermediate-level Knowledge in Interaction Design Research. ACM Trans. Comput.-Hum. Interact. 19, 23:1–23:18 (2012)

Hunicke, R., LeBlanc, M., Zubek, R.: MDA: A formal approach to game design and game research. In: Proceedings of the AAAI Workshop on Challenges in Game AI. pp. 1–12 (2004)

Järvinen, A.: Introducing Applied Ludology: Hands-on Methods for Game Studies. In: Proceedings of the DiGRA 2007 Situated Play. International Conference of the Digital Games Research Association, September 24th to 28th, 2007, pp. 134–144, Tokyo, Japan (2007)

Jetter, H.-C., Gallacher, S., Kalnikaite, V., Rogers, Y.: Suspicious Boxes and Friendly Aliens: Exploring the Physical Design of Urban Sensing Technology. In: Proceedings of the First International Conference on IoT in Urban Space, pp. 68–73. ICST (Institute for Computer Sciences, Social-Informatics and Telecommunications Engineering), ICST, Brussels, Belgium, Belgium (2014)

Kay, A.C.: A Personal Computer for Children of All Ages. In: Proceedings of the ACM Annual Conference—vol. 1. ACM, New York, NY, USA (1972)

Lee, P., Thompson, R.: Lords of Waterdeep (game). Wizards of the Coast, Seattle, WA (2012)

Lindsay, S., Taylor, N., Olivier, P.: Opportunistic Engagement by Designing on the Street. In: CHI '12 Extended Abstracts on Human Factors in Computing Systems, pp. 1709–1714. ACM, New York, NY (2012)

Lombardi, M.M.: Authentic learning for the 21st century: An overview. Educause Learn. Initiative 1, 1–12 (2007)

Löwgren, J.: Toward an Articulation of Interaction Esthetics. New Review of Hypermedia and Multimedia. 15, 129–146 (2009)

McDaldno, A.: The Quiet Year (game). Buried Without Ceremony, Pittsburgh, PA (2013)

Morozov, E.: To Save Everything, Click Here: The Folly of Technological Solutionism. PublicAffairs, New York, NY (2014)

Ratto, M.: Critical Making: Conceptual and Material Studies in Technology and Social Life. Info. Soc. 27, 252–260 (2011)

Ratto, M., Boler, M. (eds.): DIY Citizenship: Critical Making and Social Media. MIT Press, Cambridge MA (2014)

Resch, B., Summa, A., Sagl, G., Zeile, P., Exner, J.-P.: Urban Emotions - Geo-Semantic Emotion Extraction from Technical Sensors, Human Sensors and Crowdsourced Data. In: Gartner, G. Huang, H. (eds.) Progress in Location-Based Services 2014, pp. 199–212. Springer International Publishing (2015)

Rieser, M.: Locative voices and cities in crisis. Studies in Documentary Film. **6**, 175–188 (2012)

Rosenberg, U.: Agricola (game). Lookout Games, Berne, Germany (2007)

Rogers, Y.: Interaction Design Gone Wild: Striving for Wild Theory. Interactions. **18**, 58–62 (2011)

Schouten, B.A.M.: Playful Empowerment. Inaugural address May 12, 2015. Amsterdam University of Applied Sciences, Amsterdam (2015)

Schouten, B.A.M., Deen, M., Bekker, M.M.: Playful Identity in Game Design and Open Ended Play. In: Proceedings of the Homo Ludens conference, pp. 127–137. Technische Universiteit Eindhoven, Eindhoven, Netherlands (2010)

Shaffer, D.W.: How Computer Games Help Children Learn. Palgrave Macmillan, New York (2006)

Shklovski, I., Chang, M.F.: Urban Computing—Navigating Space and Context (2006)

Spinuzzi, C.: The Methodology of Participatory Design. Tech. Commun. **52**, 163–174 (2005)

Stals, S., Smyth, M., Ijsselsteijn, W.: Walking & Talking: Probing the Urban Lived Experience. In: Proceedings of the 8th Nordic Conference on Human-Computer Interaction: Fun, Fast, Foundational, pp. 737–746. ACM, New York, NY, USA (2014)

Stevens, Q.: The Ludic City: Exploring the Potential of Public Spaces. Routledge, London; New York (2007)

Stolterman, E., Wiberg, M.: Concept-Driven Interaction Design Research. Human-Computer Interaction. **25**, 95–118 (2010)

Suchman, L., Blomberg, J., Orr, J.E., Trigg, R.: Reconstructing Technologies as Social Practice. Am. Beh. Sci. **43**, 392–408 (1999)

Tan, E., Portugali, J.: The Responsive City Design Game. In: Portugali, J., Meyer, H., Stolk, E., Tan, E. (eds.) Complexity Theories of Cities Have Come of Age. pp. 369–390. Springer, Berlin, Heidelberg (2012)

Teli, M., Bordin, S., Menéndez Blanco, M., Orabona, G., De Angeli, A.: Public design of digital commons in urban places: A case study. Int. J. Hum Comput Stud. **81**, 17–30 (2015)

Townsend, A.M.: Smart Cities: Big Data, Civic Hackers, and the Quest for a New Utopia. WW Norton & Company, New York, NY (2013)

Vines, J., Clarke, R., Wright, P., McCarthy, J., Olivier, P.: Configuring Participation: On How We Involve People in Design. In: Proceedings of the SIGCHI Conference on Human Factors in Computing Systems, pp. 429–438. ACM, New York, NY, USA (2013)

Wrede, K.-J.: Carcassonne (game). Hans im Glück Verlags, München, Germany (2000)

Zimmerman, J., Forlizzi, J., Evenson, S.: Research Through Design as a Method for Interaction Design Research in HCI. In: Proceedings of the SIGCHI Conference on Human Factors in Computing Systems, pp. 493–502. ACM, New York, NY, USA (2007)

Engaging with the Smart City Through Urban Data Games

Annika Wolff, Alan-Miguel Valdez, Matthew Barker, Stephen Potter, Daniel Gooch, Emilie Giles and John Miles

Abstract This chapter will explore how gamification can be used to motivate citizens to engage with data about their city. Through two case studies, we aim to show how prompting hands-on experiences with urban data can improve data literacy and ultimately increase citizen participation in urban innovation and the co-creation of smart city apps. The first case study presents a game called 'Turing's Treasure' designed to elicit design features and data from the players for MotionMap, an interactive map that improves the planning of travel through different modes of transport around Milton Keynes, UK. The second case study describes the outcome of several creative and competitive app design sessions that have been conducted with school children in London and Milton Keynes. We conclude by discussing where we think this field is heading in the future and what additional benefits this will bring.

Keywords Gamification · Data · Co-creation · Design · Transport · Smart city · Data literacy

---

A. Wolff (✉) · A.-M. Valdez · M. Barker · S. Potter · D. Gooch · E. Giles
The Open University, Walton Hall, Milton Keynes MK7 6AA, UK
e-mail: annika.wolff@open.ac.uk

A.-M. Valdez
e-mail: a.m.valdez@open.ac.uk

M. Barker
e-mail: matthew.barker@open.ac.uk

S. Potter
e-mail: stephen.potter@open.ac.uk

D. Gooch
e-mail: daniel.gooch@open.ac.uk

E. Giles
e-mail: emilie.giles@open.ac.uk

J. Miles
University of Cambridge, Trumpington Street, Cambridge CB2 1PZ, UK
e-mail: jcm91@eng.cam.ac.uk

© Springer Science+Business Media Singapore 2017
A. Nijholt (ed.), *Playable Cities*, Gaming Media and Social Effects,
DOI 10.1007/978-981-10-1962-3_3

1   Introduction

Cities are diverse, complex and ever-changing environments. As they evolve, the physical and social infrastructure must adapt to support the increasing population. Cities must ensure that they can continue to provide an adequate supply and fair distribution of fixed resources such as energy and water, as well as effective management of the healthcare, transport and other key services. This is the responsibility of a large number of entities within the city, such as local and national government, private companies, health services and citizens. This chapter will focus on what *data* can tell us about the past, present and future state of a city and how this information can be used to adapt cities and the behaviour of the citizens who live in them to create a more sustainable future. It is increasingly recognised that this *smart city* approach to sustainable development benefits from the input of citizens towards bottom-up innovation, as a complement to top-down city-led initiatives. Citizens, being the ultimate beneficiaries, should co-create the smart technologies in order to ensure their success (Gooch et al. 2015). Citizens have better insight into the problems that they face and they are often contributors and owners of the data that drives smart city applications. The Smart City that is developed through a balance of top-down and bottom-up approaches has the possibility to create technologies that truly address the needs of the people who live and work there.

However, there are barriers to engage citizens with smart city applications and smart city design. Key amongst them is that most citizens are not sufficiently conversant with using and interpreting data to use it as a resource for learning about and modifying their behaviour and environment. In this chapter, we propose how the use of gamification can overcome some of these barriers to citizen engagement with urban data sets, moving them closer to a position where they are fully prepared to be active participants in the co-creation of smart city solutions. Through a mixture of both concrete and envisioned scenarios that have been developed as part of a smart city project called MK:Smart (www.mksmart.org) we will introduce the notion of Urban Data Games—gamified, interactive tasks using urban data sets—and explore different ways in which these might help citizens to engage more meaningfully with their environment through data and thus to change the relationship between the citizen and their smart city.

2   Data in the Smart City

Urban data is increasingly becoming a resource for innovation. It is collected through sensors, smart metres and satellite imagery. It is acquired in various ways via mobile devices, either passively using phone technology to collect data such as location, or being contributed actively by a user, for example through crowdsourcing. The data relates to a large variety of topics, including crime data, weather, energy and water consumption, transport, pollution, council services, health, social and cultural life of the city. There is information about shopping, housing, demographics and other details of the personal lives of the city inhabitants.

Some data can be delivered as live streams which provide real-time indicators of the state of the city system. Access to real-time data allows immediate adaptation to a circumstance, such as re-planning a travel route based on up-to-the-minute traffic information. Historical data can be used to understand what has happened in the past, to discover patterns and trends and to predict the future. This can be used to inform either immediate behaviour, or this can provide insight for longer term planning strategies for city design and smart technologies. For example, analysing patterns of energy consumption across a city can support utility companies to design strategies for reducing energy use during times of peak demand and to predict when these peaks will occur.

Urban data is becoming increasingly open, such that anyone—from citizens to businesses—can equally use the data as a resource both in the design and delivery of innovative smart city applications. However, much of this resource is currently under-utilised, particularly when it comes to citizens designing their own smart city solutions. Both the US and the UK have websites (data.gov and data.gov.uk, respectively) that contain large numbers of data sets that citizens can use and innovate with. However, there is currently a mismatch between the release of data and its use by citizens. At the time of writing, the UK site contains 22,759 data sets but only 378 apps are using that data. The US site contains 159,206 data sets but lists only 76 apps. Currently, the use of data is mainly by developers rather than general citizens, who usually have little awareness of open data platforms.

A number of barriers remain to the widespread adoption and use of Open Data. First, city governments need to have the institutional culture to encourage the release of data and mechanisms in place to ensure the quality and accuracy of the released data (Bertot et al. 2010; Janssen et al. 2012). Second, open data on its own has little intrinsic value; the value is created by its use by citizens. The current situation is that open data systems offer few incentives for users—there is very little promotion of open data systems to the general public, they are hard to use and finding the data you want is not straightforward (Janssen et al. 2012).

The question then is how to effectively engage citizens with data on a deep enough level that they not only understand what data can tell them about the environment in which they live and work, but can start to see the affordances of data for designing solutions to their problems. This is a critical starting point towards citizens actively using and designing with data for their own local problems. In addition to moving towards true bottom-up co-creation in smart cities, the engagement of citizens with data will help them to see the benefits of their own data contribution towards driving the products and services that will ultimately benefit them.

3 Background

Game-based principles such as goals, rewards, challenge and incremental skill-building are often used in non-game scenarios to increase engagement (Iacovides et al. 2011). This is commonly referred to as 'gamification' (Deterding et al. 2011) and is distinguished from serious games, which are games in the more conventional sense but that are used for something other than entertainment. Common examples include games for training, education and stroke rehabilitation. Kapp (2012) identifies that narrative principles are important aspects of games that provide coherence. They engage the player's imagination and are natural way for them to explore and make sense of the world (Schank 1990).

Gamification as a concept is not without its detractors who argue that the use of scoring systems as a motivator (which is only one form of gamification) can improve extrinsic motivation at the cost of intrinsic motivation (e.g. Hecker 2010; Mekler et al. 2013). Nicholson (2012) argues that "the underlying message of these criticisms of gamification is that there are more effective ways than a scoring system to engage users". However, it is generally accepted that whilst gamification may only encourage transient forms of motivation, it is heavily context dependent (Decker and Lawley 2013; Hamari et al. 2014; Schiefele et al. 1992). As such, it is worth considering what previous gamification approaches within an urban context have found.

The use of games and gamification to engage people with their urban surroundings is not a new idea. In one approach, the city streets are turned into a stage for interactive theatre in which actors masquerade as members of the public whilst a small 'audience', far from being passive, must solve clues that lead them on a journey in which they solve mysteries and discover more about the city in the process. One example is 'Accomplice the show' (http://accomplicetheshow.com/), which describes itself as 'adventure theatre' that plays out on the streets of New York. A further example was 'Uncle Roy All Around You', which combined online players with on the street audience in a challenge to find a mysterious figure known as 'Uncle Roy' (Benford et al. 2004). In both cases the clue-solving and narrative game elements are designed to prompt the players to break out of their more usual passive audience role and to participate actively as an integral part of the plot. Without the audience involvement, the play does not end and the audience is left without a resolution. An analysis of audience experience during the play revealed that many people found the experience quite disturbing, in that it elicited a feeling of paranoia. Interestingly, this emotion was regarded by audience members as being a positive part of their gamified theatre experience, contributing towards their entertainment.

Cultural games are not framed as 'theatre', but nevertheless use gamification to engage tourists with the culture of a city. Examples include City Treasure (Botturi et al. 2009), which leads children on a treasure trail around a city, using SMS messaging to mediate interaction and send clues. Goals are tailored to the age of the group doing the hunt, but effectively they lead the children into learning more about the cultural sites in the city. There is a competitive element to engage children, giving points for the types of observations they make. Also aimed at children, O'Munaciedd

(La Guardia et al. 2012) is a treasure hunt in which children explore a city by trying to find a character called O'Munaciedd and grab his hat. Their route and clues are given on a hand-held device. In answering the clues, children also learn something about the culture of the city. In REXPlorer (Ballagas et al. 2008), tourists are wizards, who can cast spell by waving their mobile device in a certain way. By casting a spell, the tourist gets to hear stories related to their current location, as well as being sent on quests. As tourists undertake quests they also take photos. At the end of the game the story of their travels is turned into a blog. The Lost Lab of Professor Millenium (Kuikkaniemi et al. 2014) combined augmented reality navigation techniques with a series of puzzles to prompt young learners to find out more about Helsinki. Hello Lamp Post (http://www.hellolamppost.co.uk/) was a city-wide project in Bristol, UK that ran for two months during 2013. It was designed to prompt playful engagement not with cultural sites, but instead with typically overlooked city locations and street furniture, through the sharing of stories which were submitted via text message and associated with objects via a unique id. This project resulted in 20,000 contributed messages—a success that was repeated when the same approach was used in Austin, Texas, thus demonstrating the potential of a playful approach to engage a large number of people with their environment.

An alternative use of gamification within a city is in bringing communities together. For example, the ZWERM system (Coenen et al. 2013, 2014) was designed to increase the sense of community in two neighbourhoods in Ghent. The two neighbourhoods were competing to gather the most points. Citizens could gain points by 'checking in' by placing an RFID card on a reader on a hollow tree in a public space within the neighbourhood. More points were earned if two players checked-in together (encouraging players to talk to their neighbours). The ZWERM system was relatively successful—277 people took part and the results of the study indicate that the game was successful at increasing social capital. However, whilst the authors present a description of the living labs process used to generate the game, they offer no insight into the robustness of the methodology or its strengths and weaknesses. In City Game (Games: at field of view 2012), players use Lego-like blocks and follow a set of rules to collaboratively develop a city plan. The goal for City Game is to provide insight to city planners about the needs of different communities, as represented by the players and for the players to come to understand more about the conflicts and constraints of city planning. However, City Game is dissociated from the real-life urban context. Another example of bringing communities together through play is the project 'Hippokampos in the Grey Matter' (2012) by the women's art and technology group MzTEK, in collaboration with designer Pollie Barden. This saw women in Athens participate in a workshop in which they together designed a game that was then played by members of the public the following day. The aim was for people to reflect on their physical environment whilst playing, but also to encourage more women to engage with creative technology practices through the involvement of MzTEK.

Through the above, we have provided some examples of how games can engage citizens with the urban environment. However, these commonly lack a focus on empowering the players to understand better how to change and adapt their environment, or to change their own behaviour to better fit with the environment. We propose that engaging citizens with data about their environment can help in this transition. If citizens understand what the data tells them about the past, the present and how this can predict the future, then citizens gain insight into both what needs to change and potentially how to bring that change about. In the exhibition 'a conversation between trees' (Jacobs et al. 2013) art was used to prompt deeper reflection on climate change data. Whilst not framed as a 'game', the experience was designed, much like a narrative, to elicit an emotional response from the audience, such that they would reach better clarity about their own perspective on the topic which would be more likely to lead to behaviour change. Another example, 'Summer School Data Explorers' (2015), combined data with art through workshops led by technology education company Codasign. School children in transition year (about to begin secondary education) collected their own environmental data using sensors and Raspberry Pis within the Queen Elizabeth Olympic Park, London, UK and used their artistic skills to produce visual analyses of their data. This was turned into a site-specific piece of art by designer Stefanie Posavec, reflecting the findings as a group. In these cases, interaction with the environment is not only direct but is also mediated through the lens of the data and the interpretations (both analytical and artistic) that are made about it.

Urban policy making is increasingly influenced by data, as are patterns of behaviour. But technologist's conceptions of data as neutral and objective are fictions; data are inherently partial, selective and representative and the criteria used in their capture have consequence (Kitchin 2014a). Thus, the smart city agenda can become an arena of collaborative argumentation, helping to identify what the issues are, how they are understood and what possibilities there are for action (drawing on Healey (1997) about participatory discursive democracy). Such an approach can be achieved in a variety of ways, and opens up a role for urban data games.

Games can play an important role in smart city developments, but that role depends on what type of smart city is envisaged (Kitchin 2014b). If a smart city is one where businesses and government organisations develop smart city systems for their purposes of efficient management and control, then gaming elements may be used to ensure the understanding of citizens. A more interactive approach is to use serious games to educate citizens' perspectives on information to inform their choices as consumers. A further stage is where, through gamified approaches, citizens take an active role in managing and in co-creating the smart systems themselves and collect and use data which is meaningful to them. Gamification in this latter context is not just about educating citizens about using and valuing smart city big data systems, but in them being empowered to have genuine input into the design of those systems. This is about using smart city systems for democratisation. These approaches to urban games are not mutually exclusive, and elements of each have been explored in the context of a smart city project taking place in Milton Keynes, a town with a population of 260,000 located in the Southeast of England.

The vision behind the MK:Smart programme is built around a smart city concept that seeks city efficiency through a focus on using ICT for democratisation.

MK:Smart is a £16 million smart city initiative, co-funded by the Higher Education Council for England, taking place in Milton Keynes between 2014 and 2017 (MK:Smart 2016).

4 Urban Data Games

The remainder of this chapter will explore two quite different scenarios in which gamification is used to prompt user engagement with urban data in the co-creation of smart city technologies. The first is in the co-creation of MotionMap to create a shared resource to improve travel planning around Milton Keynes, UK. The second is in a gamified task in which young students design smart city applications as part of an initiative to improve data literacy of school leavers in the U.K.

We define these as Urban Data Games, which use one or more aspects of gaming such as narrative, challenge or reward as a mechanism to prompt engagement with an urban data set in the co-creation of smart city technologies. Within an Urban Data Game, the gamification principles are designed to encourage the user to engage in behaviours that are closely aligned with the desired outcomes. For example, if the goal is to encourage users to engage with a smart city app, or to contribute data to an app, a gamified approach could be to set a goal that is only achievable through these behaviours. In designing an Urban Data Game it is therefore important to be clear about (1) what are the intended outcomes, independent of the game (2) what will motivate the user towards those outcomes. Gaming elements should be most focused on activities that users would not naturally be motivated to undertake, either those that they find difficult or those that they find mundane.

4.1 MotionMap

Within the MK:Smart programme, a smart transport initiative called MotionMap is currently (2016) in development. MotionMap is intended to demonstrate the concept of 'Cloud Enabled Mobility', an app that connects users with information and other cloud-based services to enable informed transport decisions. A large number of mobility information apps already exist, varying from driver GPS guidance to parking availability, bus and rail information, bicycle sharing and taxi booking, among many others. Many of these are isolated supplier-led apps, some seek to integrate travel information across a city (e.g. Citymapper), but they often do not provide real-time data and are a centralised approach that provides data to users that providers consider they should have, rather than information that users have had a role in specifying as meeting their needs.

The approach in MotionMap will go beyond the information provided by existing travel apps, in that it will continuously describe the real-time movements of people and vehicles across the city, providing embedded timetables and estimates of conges-

Fig. 1 A MotionMap
mock-up

tion and crowd density in different parts of the city. It is also envisaged as a platform whose design and operation incorporates the needs of its community. It will also be a platform on which community groups, businesses and other organisations can develop their own applications. Figure 1 shows a mock-up of MotionMap.

To bridge the gap between this abstract concept and the specifics of its implementation, a key desire in MK:Smart is to engage citizens in the co-creation of solutions using the MotionMap platform. The intention is also to design the system to allow users to actively contribute relevant data in order to increase its value to all. Some of the information required for smart transport can be generated automatically, through feeds from existing databases and traffic sensors, etc. However, the system provides greater legitimacy and informational richness if users contribute information that is relevant to them and their peers (for example, reporting traffic jams, rough surfaces and faults on cycleways or tardy bus services).

Citizen workshops have taken place as part of the MK:Smart programme, including the use of illustrative non-functional prototypes and mock-ups of MotionMap to foster discussions in user workshops. Here participants were asked to discuss the features of MotionMap that they would value. The semi-structured discussions prompted by the prototypes were messy, with participants talking about their concerns, existing practices and ways in which smarter transport information might improve their quality of life or make their transport more sustainable. This information was relayed to the software developers, shaping the features of the app. As the functionality gradually increased, new prototypes and new scenarios could be introduced into the conversation. This iterative process continued until there was sufficient functionality to transition from the workshops into an approach based on urban games.

4.1.1 Urban Games for Smart Transport: The Case of Waze

The urban gaming strategy pursued by the MotionMap team was based on the study of similar smart transport projects, deployed on a commercial scale, and which demonstrated that games could be used to foster user engagement and facilitate user-centric design. Inspiration was drawn from several games and serious gaming applications [including Open Street Map's mapping parties (Hristova et al. 2013), Citymapper's provision of estimates for jetpack users, Strava's virtual 'King of the Mountain' competitions for cycle users (Smith 2015)], but particularly from a motorist app, Waze.

Waze was founded in 2008 in Israel. Its flagship product is a GPS-based app for driving navigation with turn-by-turn instructions that adjust to traffic in real time, suggesting alternative routes for users to get around congested areas. This has subsequently been developed to provide other user benefits, including ridesharing matches. Data is generated entirely by Waze users themselves, meaning that a community can make it effective without linking to other GPS information (Waze 2016). It is an entirely self-contained user data generation and sharing system. This crowd-sourced design does not conform to the usual smart transport model built around centralised infrastructure instrumentation; Waze is smart transport of a different type, and a very successful one. Waze has expanded globally to have a user base of approximately 50 million people worldwide. Its approach attracted the attention of Google who, in June 2013 bought Waze for $966 million.

The challenge identified by Waze's founders was that of getting to critical mass. Widespread adoption will not take place unless the app can provide something valuable to early users. However, offering a service of actual practical value can be difficult in the early stages of diffusion of an urban sensing system. In response, the development model followed by Waze (and adopted by MotionMap) includes gaming as a crucial feature. This model consists of four stages

- Build
- Play
- Provide value
- Achieve critical mass

The first stage involves building a minimal base system (e.g. providing a map of the basic road grid). At this stage, it is not completely functional, but there is enough for early adopters to play with, and to demonstrate the application's potential. Thus, ludic value can be delivered through the inclusion of actual game-like elements, but also by making a system that makes user contributions rewarding and that is fun to tinker with. Eisnor (2011) summarised this as " You can't deliver a lot of value, but people see that it is working and growing". If this 'playful' stage attracts the attention of a sufficient number of potential users, they act as value co-creators. The system can then grow to the point where it can provide practical value. It is in this stage of the approach that the service starts to become dependable and achieves critical mass.

When Waze is introduced in a new city and insufficient data is available, users driving on an unmapped road can transform the application into a real-world version of a video game with their on-screen avatar transforming into a 'Pac-Man' eating power pellets and other rewards whilst routing information is automatically collected in the background. Whilst more seriously-minded users are free to ignore this game-like aspect of the application, Waze reports that 8–12 % of the users will go out of their way to earn a virtual cupcake. Through the increased efforts of those 12 % of gamers, sufficient data can be collected to deliver practical functionality to the complete user base. Waze does not only use gamification as a development tool, but in the design of its interfaces with users, with engaging playful features that support the overall user-focused approach.

4.1.2 MotionMap Games

The urban gaming strategy for MotionMap relied on playful value to foster early engagement, before practical value could be delivered. Early functionality of MotionMap had a very limited coverage, but did include traffic sensing for a limited number of roads, there were a few car parking areas that had sensors indicating if they were occupied or not, and real-time bus scheduling information was available for some stops. This patchy sensing data was made available by several teams of technical developers as they deployed proofs of concept for different monitoring solutions. These could not offer practical value, but they provided enough glimpses of the city for a virtual treasure hunt to be designed. This was a small-scale game, linked to the Freshers' Week for full-time Open University Ph.D. students held on the OU campus in Milton Keynes in October 2015. Figure 2 shows some publicity material for this hunt.

This virtual treasure hunt was entitled *Turing's Treasure*. It weaved the limited data available into a story about Alan Turing, who was a computing pioneer based on the secret Second World War code-breaking centre at Bletchley Park. Turing actually buried and subsequently lost a cache of silver somewhere in Milton Keynes, so this story was used to provide play value for a very early stage of MotionMap that provided little practical value at all. MotionMap made this limited data from sensors available visually, but also in numerical format. The game used this numerical information to provide clues allowing the gamers to solve a cryptogram identifying the virtual location of the treasure.

Fig. 2 Publicity for Turing's Treasure

The treasure hunt had a duration of one week, with daily clues and challenges which required interaction with the transport data made available through the online prototype MotionMap. Figure 3 shows the winners receiving their prizes. The interaction of the players with the game was tracked through online metrics, through their responses to the challenges and, most importantly, by inviting direct feedback. The game encouraged sustained interaction, which created awareness of its capabilities and limitations, and reflection about potential practical applications. The literature on interactive systems design suggests that prototypes can be used as 'straw men' to elicit customer input during requirements gathering, providing a target point for discussion and a target for criticism (Rudd et al. 1996). In the case of *Turing's Treasure*, the introduction of playful elements into the prototype made it possible to sustain this process for a week, in contrast to more traditional user workshops where the duration is limited to a few hours.

Some of the feedback produced by this process was related to minor technical issues and bug reports were raised by users, but some users sent detailed letters with suggestions and features they would like to see in future iterations of the game, or even volunteered to contribute to development efforts. Whilst users do not expect to use the current version of MotionMap for practical purposes, they described features and applications they would like to see. Interacting with the game contributed to envisioning of the complete product, and to proposals about features that would be of interest.

Insights from the iterative design process facilitated by prototypes and games are contributing to a gradual re-definition of the scope of cloud-enabled mobility. MotionMap was initially designed with car users in mind. Features most emphasised in the game and in the mock-ups used in the workshops targeted drivers (for example, with questions in the virtual treasure hunt involving the use of parking sensors). However, users were more interested in features related to alternative forms of transport.

The walking and cycling features of MotionMap played a limited role in the game, but attracted the attention of users. Whilst car users in Milton Keynes were found to

Fig. 3 Awarding the prizes to
the two games winners

be generally satisfied, pedestrians and cyclists declared that some smart support
would be welcome. Users were interested in maps and navigation instructions, but
also in road reports generated through automated and crowdsourcing methods. For
example, the accelerometers that are integrated with some smartphones can be used
to produce automatic reports regarding bad surfaces, steep gradients and unexpected
sharp bends in cycleways. There was interest in complementing this automatically
generated data by crowdsourcing information about hazards like glass on the cycleways, areas with insufficient illumination, or flooded underpasses. By providing
advance notice of the hazards, users can plan alternative routes, or ensure that they
are riding at a safe speed when they approach problematic sections of the roads.

Public transport users were also enthusiastic about MotionMap. Through their
interactions with the game and the prototypes they envisioned and requested additional functions that were not foreseen originally, but that could be supported with
the smart city infrastructure. Particularly, they were interested in using the sensing
systems for providing their own crowd-sourced, real-time reports about the location
of buses. One key aspect to emerge from the workshops and discussions was that
user crowd-sourced monitoring activity would make transport providers more
accountable. This, in effect, would counteract the existing disempowering relationship between bus users and service providers. At the moment, transport providers
are the sole gatekeepers of information, and there is limited accountability for delays
or interruptions to the service. Shifting this relationship so that users become information generators and holders is valued and would be a radical step, which will be
exploring in the next iteration of the game.

4.1.3 Lessons Learned

By playing a game, when only limited MotionMap functionality had been developed narrowed the scope of this game, but made it possible to integrate users into the design process early, before major design decisions had been set in stone. This early inclusion through the gamification of user testing provided several lessons

1. Some lessons were purely technical. The game tested the smart city infrastructure, and its capability to keep MotionMap and all the feeds running and accessible to several users on a variety of systems and browsers;
2. It tested the suitability of using a game as a vehicle for creating awareness and fostering user engagement;
3. The game provided feedback about usability, and prompted reflection not only about the MotionMap, but also about its potential applications. Because this feedback was received early in the development process, it could have a meaningful impact on the final version of the MotionMap app.

These results fed into an MK:Smart workshop considering how gaming approaches could be further used for MotionMap and also applied in the development of the MK:Smart's other workstreams. One key conclusion was that the role of gaming changes as a smart city initiative develops. There should not be a single function of urban games but, as in Waze and other examples, there need to be layers of gaming that

(a) Provide early stage developmental feedback
(b) Helps users to envision and build their own applications based on the product
(c) Encourages early stage user engagement
(d) Supports the development of a community of champions
(e) Provides engagement with the final product
(f) Provides rewards to users who engage in socially beneficial behaviours.

These different styles and aspects of serious gaming and gamification are not necessarily sequential, but need to be present for the needs of different users at different times. For MotionMap, serious gaming activities initially targeted points (a) and (b), using urban games as part of a long-term strategy of iterative prototype-based co-creation. In order to engage a growing number of citizens into this process, we are now moving to focus on functions (c) and (d) with gamification elements to be incorporated in the design of the app features.

4.2 Urban Data School

There are a number of barriers to citizen participation in the planning and design of sustainable 'smart city' solutions. One key limiting factor is that the majority of citizens lack practical experience in handling and using complex data sets in the design of physical products and services. Thus, currently, the expert use of data is the preserve of businesses and government organisations, who have more expertise to draw on for utilising this data. We need to find ways to engage public with the urban design process and at the same time improve their knowledge and conceptual understanding of the relationship between data produced by people as they live and move around the city and the physical urban environment they inhabit.

The Urban Data School (UDS) is an initiative aimed at delivering content around complex, urban data sets to improve data literacy of school leavers. A number of creative app design sessions have been conducted as part of the UDS. These are designed to give students experience of imagining how data can be used as part of smart city innovation. In the app session the students are set up into competitive teams and given a challenge to design the best app which uses some collected form of data and which can solve a problem in their local area. They are the only activities in the Urban Data School which do not use real data sets—instead the students must use their imagination to think what sort of data they could use, how it would be collected and what this might be able to tell them. Students are prompted to design non-standard visualisations of this data into their app design. The appathon is a form of event-based learning, based on principles of challenge-based learning (Johnson et al. 2009) which puts people under time pressure. Often a collaborative learning experience, participants are provided with some information and goals prior to the event in order to prepare, i.e. organise a team, define roles for members based on expertise and identify and address skills gaps. Thus, the appathon aims to use game-based principles of challenge and reward to motivate the students to perform well in the task, despite many of the concepts around smart data, complex data and visualisations being new to them.

Appathon 1 Occurred during a regular classroom session in a Milton Keynes school. A total of 17, year 9 (age 13–14) students, in advanced science classes, participated in three sessions over a three-week period. The first two sessions were designed around the use of complex data obtained from the wider MK:Smart project. These sessions encouraged the students to explore home energy consumption and solar generation of households in Milton Keynes by analysing smart metre data and an aerial photography map of the city, visualising potential solar energy production for the different rooftops. The final session was the app design session, which occurred during a 2.5 h classroom session with no break. Students were divided into four groups, each of which produced a unique app design.

Appathon 2 Occurred as part of an initiative that provides additional learning opportunities for young students from several schools, in the form of weekend tutorial and workshop events. In this session a total of 38 students, aged 15–16 were given a tutorial on big data and how it is used in smart city applications. Following this, the students took part in one of two app design sessions, each of which took 1 h and 25 min. There were 18 students in the first session, who divided themselves into 5 groups and 20 in the second session who divided themselves into 3 larger groups.

4.2.1 Session Formats

Both appathon's began with verbal and written instructions for the participants. In both sessions students were instructed to design an app to tackle an urban problem. They were told it was a timed session, that they should work in groups and that at the end of the session they would pitch their idea. The winning group would be chosen by a panel based on the following criteria: (a) Quality of pitch, (b) Technical feasibility of the app, (c) Creativity of design, (d) Use of data and data visualisations and (e) Innovation. These criteria were designed to get the students thinking broadly and creatively about how they could design an app which would provide intelligence with real impact. Students were also provided with a list of prompts, which encouraged them to think about: 1. Who were the app users? 2. What would the interface look like? 3. Why would people use the app? 4. What data would be used and how would it be visualised? 5. How would people use the app?

The hand-out given to the groups in Appathon 2 included additional instructions to sketch the interface, as well as an example design. They were also given four additional real-life examples of urban apps based on data. In Appathon 1 (the earlier session) the groups were also asked to sketch the interface, but this instruction was given verbally.

Students were provided with flipchart paper and coloured markers. In Appathon 1, the project teams were supported both by the regular classroom teacher and two researchers from the UDS initiative. In Appathon 2, the groups were supported both by helpers who often attended the weekend sessions and were known to the students, and three UDS researchers.

The appathon challenges were designed foremost as an opportunity for students to learn more about designing smart city apps, gain better understanding of using data within design and to develop critical thinking skills. Therefore, it was necessary to offer students appropriate support to achieve this. It was also an opportunity for us to assess the competencies of the students. Consequently, it was decided that one researcher who was present at both sessions would circulate between the teams, offering support over the course of the task. They provided the same set-prompts to workgroups in both appathons, focusing on the use of data as this was key to the theme of the appathon and was identified as the most unfamiliar aspect of the task for the students. The groups were given 10–15 min to generate their initial ideas. The researcher then visited each group in turn to listen to and support their ideas and to suggest improvements. This was intended to encourage the participants to focus specifically on the use of data and the analysis of this data within their app. All groups had, at this point, identified what data they would collect and how this would drive their app's functions. The researcher prompted each group to think about how a 'smart city' analysis *across* their data sets could give their app broader scope and more powerful functionality. The researcher gave examples related to the student's own proposed data. In cases where a group was already proposing a solution in which collected data was analysed, the researcher prompted them instead to think of more novel ways to collect data. Towards the end of each session, the students were encouraged to finalise their pitch. They presented their ideas and a winner was chosen using the criteria stated above.

5 Findings

It was clear from observing the challenges that the students were engaged with the task. Each group worked together and managed to produce and present a unique app design within the time limit. Across both the appathons, with both the younger and older age group, students needed the most support and prompting in thinking about how the data they proposed to collect could be analysed to yield the most insight across the geographical region or population represented through the collective data. Whilst the app designs were never intended to be working products, it was apparent that students were thinking in terms of the brief and the majority of designs were both feasible and applicable to a real-life problem that they had identified. The app designs consisted of the following (appathon number is shown in brackets):

1. Identify Wildlife—identify wildlife from an uploaded photo (1)
2. Walking Wardrobe—show trending clothing items in the local area (1)
3. Barcode Scanner—scan your recycling and earn points (1)
4. Baby Steps—timely advice for mothers on when to tend to their baby (1)
5. Oldies—social media app for older people to connect with each other (2)
6. Hotspot—show busyness of nearby tube stations
7. Y2K Youth—find events for teenagers (2)
8. Leisure Time—meet like-minded people in your area within hours (2)
9. Taxis on Demand—match people to a nearby taxi with appropriate number of seats (2)
10. Guide Me—find safe route for people based on declared disability (2)
11. Grocery Genie—keep track of what people buy (2)
12. NHS—for making doctors appointments (2)

The Hotspot app is shown in Fig. 4. Since the appathon revealed that students had most difficulty in aspects of the challenge related to planning analyses across their 'collected' data, some new research is being planned which will focus on whether a more tangible, playful, interaction with data can support students better in this activity than the purely 'mind-based' planning within the appathon task. A tangible map will be created upon which students can plan where and how to collect different types of data and get feedback in real time. The aim is to framework a learning experience which prompts pupils to move away from considering collected data as simply a localised data source, towards considering it as a spatially located source with a temporal dimension. It is hoped that ultimately this will lead learners to understand how multiple data points are related spatially and temporally and what insights can be found by utilising this information.

Fig. 4 The Hotspot app
design

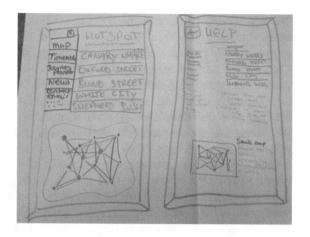

Finally, future work within the Urban Data School will investigate the use of alternative urban data games for engaging the students. One of these will be an *Eco-puzzle* which would be delivered as an 'escape the room' type challenge. In this puzzle participants would be provided with multiple, partially simulated data sets where some of the data provides clues to a recent disaster. The goal is to find out what has happened in the shortest time and present a compelling justification, backed by data visualisation and analysis. Participants must query the data sets and narrow down the time frame for an event. Data will include sensor readings that show an anomaly, simulated Twitter data of some eye-witness accounts, traffic data showing travel disruption, police reports or air pollution measurements. The goal is to apply data interpretation, analysis and visualisation skills to first identify an anomaly in one of the data sets which gives a narrative *setting* (a time and place) for the event. This is used to frame further queries across other data sets and build evidence about what occurred. Hints can be given, as there is a single, verifiable, solution but the sources of data used to evidence it can differ. Therefore, the competition may be judged on fastest or best solution, also taking into account the number of hints used.

6   Discussion and Conclusions

The city is perhaps the arena where data has its greatest utility. In places where there are high concentrations of people, information becomes richer and more prevalent. In parallel, the potential for the information to be used for social, political and environmental interventions is greater. The projects outlined above are uncovering how playing with data in the city can make the raw information a fun and meaningful resource for design and bring data handling to a larger audience.

Over the course of this chapter, we have overviewed a number of projects that handle the concepts of data and gamification in the city. Two projects, the *Turings' Treasure Hunt* and the *Urban Data School* have looked to combine these two factors.

These two projects make the participants confront data in the city from different perspectives and at different levels of abstraction. The MotionMap asks participants to be present in the city and engage with data as physical agents, 'on the ground'. The urban data school asks the participants to take a step back from the action and consider the state of the data in the city as an overseer. In spite of the difference in their approaches, they both lead to the participants reflecting upon the prosperity of their city and their fellow citizens and thinking creatively about how they can exploit the information to solve problems. The treasure hunt challenge made the participants reflect through experience. They were made to consider how the physical features of their city and their movement across it could be encoded into data and used as a resource for design as they explored the city themselves. The urban data school, meanwhile, took the participants on a journey of more internal exploration. They were made to draw upon their conceptual understanding of the city's systems and processes and reflect upon how they can use data to solve its problems without engaging with it physically.

In both cases, the data is made to seem like more than a crude set of numbers. It is given a colour and form through its relationship, with the physical environment and the pixels of an app. The gamified narrative, meanwhile, empowers the participants and makes them feel like they can have an impact upon real-world problems. This can have benefits where bottom-up design is concerned. It appears that by setting the participants against each other as competitors they are motivated to engage with low fidelity technology designs like Waze and the MotionMap for longer periods of time and have the confidence to voluntarily contribute their opinions on how the systems can be improved. In the case of the Urban Data School, the competition format makes the participants more willing to think abstractly about how different sources of information can be pulled together across a geographical region to provide insights on a city issue.

6.1 Future Work

We have only touched the surface where urban data games are concerned and there is plenty of scope for future research in this area. One of the key outcomes of urban data games is education. The engagement of the participants with data sources clearly has the potential to be a source of learning, in terms of creativity, real-world problems and data literacy itself. Gamification could be a great tool for overcoming the attrition rates that can befall other number-based topics of study (Seymour 1992). However, as discussed, it is important to not let the narrative of the game dilute the subject of study to the point where the participants are no longer getting any of the core learning. There is a specific risk that designers might be tempted to make the learning experience too superficial in the context of urban data games because of the scale and complexity of the problems that are being confronted. Future research should establish how many layers of narrative can be added to the learning experience before it is tarnished.

Another key area for future research is to try and better understand the how to engage different target audiences. The MotionMap and appathon designs were tested on specific social groups (Ph.D. students and school children), who may be more motivated to engage with technology and learning tasks than others. It would be beneficial to start building a framework to outline what kinds of data games should be used in different situations for the broader spectrum of society. One of our immediate objectives is to create urban data games that are targeted at specific groups who play a key role in the running of the city (e.g. bus drivers, policemen, councillors). These games are intended as a stimulus for the ideation of new services. Through these design sessions we aim to gain a first insight into whether urban data games can solve some more obscure city problems and engage *harder to reach* social groups. Ultimately, by tapping into the intrinsic human need to play, compete and learn we hope to cross social boundaries and bring about positive, systemic change in the areas where traditional forms of citizen engagement have failed to do so.

I have wrought my simple plan
If I give one hour of joy
To the boy who's half a man,
Or the man who's half a boy.
(Arthur Conan Doyle)

References

Ballagas, R., Kuntze, A., Walz, S. P.: Gaming tourism: lessons from evaluating rexplorer, a pervasive game for tourists. In: Pervasive Computing, pp. 244–261. Springer, Heidelberg (2008)

Benford, S., Flintham, M., Drozd, A., Anastasi, R., Rowland, D., Tandavanitj, N., Sutton, J.: Uncle Roy all around you: implicating the city in a location-based performance. Proc. Adv. Comput. Entertainment **21**, 47 (2004)

Bertot, J.C., Jaeger, P.T., Grimes, J.M.: Using ICTs to create a culture of transparency: e-government and social media as openness and anti-corruption tools for societies. Gov. Inf. Q. **27**(3), 264–271 (2010)

Botturi, L., Inversini, A., Di Maria, A.: The city treasure mobile games for learning cultural heritage. In: Proceedings of Museums and the Web (2009)

Codasign, Posavec, S.: Summer school data explorers. Available at http://codasign.com/2015/10/data-and-art-in-the-olympic-park/ (2015). Accessed 22nd Feb 2016

Coenen, T., Mechant, P., Laureyssens, T., Claeys, L., Criel, J.: ZWERM: stimulating urban neighborhood self-organisation through gamification. In: Proceedings of the International Conference on Using ICT, Social Media and Mobile Technologies to Foster Self-Organisation in Urban and Neighbourhood Governance, Delft University of Technology (2013)

Coenen, T., van der Graaf, S., Walravens, N.: Firing up the city-a smart city living lab methodology. Interdisc. Stud. J. **3**(4), 118–128 (2014)

Decker, A., Lawley, E. L.: Life's a game and the game of life: how making a game out of it can change student behavior. In: Proceeding of the 44th ACM Technical symposium on Computer science education, ACM, pp. 233–238 (2013). doi:10.1145/2445196.2445269

Deterding, S., Dixon, D., Khaled, R., Nacke, L.: From game design elements to gamefulness: defining gamification. In: Proceedings of the 15th International Academic MindTrek Conference: Envisioning Future Media Environments, ACM, pp. 9–15 (2011)

Eisnor, D.: Game mechanics and location based services: crossing the LBS chasm. https://www.youtube.com/watch?v=5_xEUjSpu6g (2011). Accessed 16th Feb 2016

Games: At field of view. http://docs.fieldsofview.in/public/fov_games_booklet_8Jun2012.pdf (2012). Accessed 22nd Feb 2016

Gooch, D., Wolff, A., Kortuem, G., Brown, R.: Reimagining the role of citizens in smart city projects. In: Proceedings of the 2015 ACM International Joint Conference on Pervasive and Ubiquitous Computing and Proceedings of the 2015 ACM International Symposium on Wearable Computers, ACM, pp. 1587–1594, (2015). doi:10.1145/2800835.2801622

Hamari, J., Koivisto, J., Sarsa, H.: Does gamification work? A literature review of empirical studies on gamification. In: Paper presented at the 47th Hawaii International Conference on System Sciences, IEEE, pp. 3025–3034 (2014). doi:10.1109/HICSS.2014.377

Healy, P.: Collaborative planning: shaping places in fragmented societies. Macmillan Press (1997)

Hecker, C.: Achievements Considered Harmful? In: Game Developers Conference (2010)

Hristova, D., Quattrone, G., Mashhadi, A. J., Capra, L.: The Life of the Party: Impact of Social Mapping in OpenStreetMap. In: ICWSM (2013)

Iacovides, I., Aczel, J., Scanlon, E., Taylor, J., Woods, W.: Motivation, engagement and learning through digital games. Int. J. Virtual Pers. Learn. Environ. 2(2), 1–16 (2011)

Jacobs, R., Benford, S., Selby, M., Golembewski, M., Price, D., Giannachi, G.: A conversation between trees: what data feels like in the forest. In: Proceedings of the SIGCHI Conference on Human Factors in Computing Systems, ACM, pp. 129–138 (2013)

Janssen, M., Charalabidis, Y., Zuiderwijk, A.: Benefits, adoption barriers and myths of open data and open government. Inf. Syst. Manag. 29(4), 258–268 (2012)

Johnson, L.F., Smith, R.S., Smythe, J.T., Varon, R.K.: Challenge-Based Learning: An Approach for Our Time. New Media Consortium (2009)

Kapp, K. M.: The gamification of learning and instruction: game-based methods and strategies for training and education. Wiley (2012)

Kitchin, R.: The data revolution: big data, open data, data infrastructures and their consequences. Sage (2014a)

Kitchin, R.: The real-time city? Big data and smart urbanism. GeoJournal 79(1), 1–14 (2014b)

Kuikkaniemi, K., Lucero, A., Orso, V., Jacucci, G., Turpeinen, M.: Lost lab of professor millennium: creating a pervasive adventure with augmented reality-based guidance. In: Proceedings of the 11th Conference on Advances in Computer Entertainment Technology, ACM, p. 1 (2014)

La Guardia, D., Arrigo, M., Di Giuseppe, O.: A location-based serious game to learn about the culture. Proc. FOE (2012)

Mekler, E.D., Brühlmann, F., Opwis, K., Tuch, A.N.: Disassembling gamification: the effects of points and meaning on user motivation and performance. In: CHI'13 extended abstracts on human factors in computing systems, ACM, 1137–1142 (2013). doi:10.1145/2468356.2468559

MK:Smart. http://www.mksmart.org/ (2016)

MzTEK, Barden, P. (2012) Hippokampos in the grey matter. Available at http://www.emiliegil es.co.uk/Hippokampos-in-the-Grey-Matter. Accessed 8th March 2016

Nicholson, S.: A user-centered theoretical framework for meaningful gamification. Proc. GLS 8 (2012)

Rudd, J., Stern, K., Isensee, S.: Low versus high-fidelity prototyping debate. Interactions 3(1), 76–85 (1996)

Schank, R.C.: Tell Me a Story: A New Look at Real and Artificial Memory. Charles Scribner's Sons (1990)

Schiefele, U., Krapp, A., Winteler, A.: Interest as a predictor of academic achievement: a meta-analysis of research. In: Renninger, K., Hidi, S., Krapp, A. (eds.) The Role of Interest in Learning and Development. Lawrence Erlbaum Associates, Hillsdale, NJ (1992)

Seymour, E.: "The Problem Iceberg" in science, mathematics, and engineering education: student explanations for high attrition rates. J. Coll. Sci. Teach. 21(4), 230–238 (1992)

Smith, W.R.: Communication, sportsmanship, and negotiating ethical conduct on the digital playing field. Commun. Sport (2015)

Waze: About us. https://www.waze.com/about (2016). Accessed 16th Feb 2016

Size and Shape of the Playing Field: Research Through Game Design Approach

Viktor Bedö

Abstract This chapter explores how to set up a game design-based research work-shop format following the *research through design* paradigm. It elaborates on the 'Size and Shape of the Playing Field' workshop series initiated by the independent research lab Tacit Dimension to explore the following research question: from what kind of geodata is it possible to generate insight on the experience of walking on the street? The workshops are generating insight by involving participants with inter-disciplinary backgrounds in an urban game design-related activity and reflection process. This chapter provides an account of the first two workshop sessions held in Berlin and Amsterdam in 2016.

Keywords Urban game design · Geodata · Research through design · Design research · Implicit knowledge · Prototyping

1 Introduction

There are many ways in which games and playfulness drive explorative urban design and generate insight about the urban condition. Playfulness can invite us to interact with experimental setups like lamp posts that can listen to our story and retell it to pedestrians in other locations. A growing number of pioneering projects imbued with a maker or hacker attitude and a high affinity for technological experimentation draw upon playfulness to conciliate disruptive urban designs with everyday urban routines (see Nijholt 2015). The majority of the playful systems installed in urban spaces create interactions using sensors (like touch, voice, heat, etc.) and actuators (such as small servo motors, light sources, etc.). The project *Hello Lamp Post*, for example, lets pedestrians talk to lamp posts and other urban furniture, so that the system records their stories and retells them to pedestrians somewhere else in the

V. Bedö (✉)
Emserstraße 65, 12051 Berlin, Germany
e-mail: viktor@tacitdimension.com

© Springer Science+Business Media Singapore 2017
A. Nijholt (ed.), *Playable Cities*, Gaming Media and Social Effects,
DOI 10.1007/978-981-10-1962-3_4

city through another piece of street furniture.[1] Gamifying approaches introduce a playful layer to everyday urban life and can be employed to crowdsource data, like in the case of the smartphone-based game *Ingress*.[2] More specifically street games and urban games—mostly with an explicit ruleset, a beginning, and an end—are predominantly physical games that incorporate the built urban environment and urban layout, and build on social, political, and architectural affordances of urban sites. They temporarily amalgamate the rules of everyday urban life with new or even disruptive interactions to create coherent and meaningful experiences. Such games can be used as mapping tools to foster understanding of the physical, social, or even political aspects of urban sites.[3] Urban games can also serve as prototypes for in situ testing of urban design concepts or location sensitive technologies (Bedö, forthcoming 2017). In all of these ways games and playfulness have a role in research for and testing of future urban designs and interactions that drive urban innovation. This chapter focuses on the activity of gaming and game design itself as knowledge generating activities in the framework of the *research through design* approach. This means that the primary aim of a design process is leveraging and negotiating embodied knowledge in a hands-on and contextualised way in order to generate knowledge for a more general research question. The main benefits of specifically drawing upon urban game-related design activities compared to other research methods is that the confined spatial and temporal extension of a game allows for a lab-like setting of variables and that, through high involvement, it enables a fast buildup of expertise from players' side. The chapter outlines how a game design-based workshop is set up to address the following research question: from what kind of geodata is it possible to generate insight on the experience of walking the street?

Technically geodata is any kind of data with geographical coordinates attached to it. Practically, geodata could be anything from data about the energy consumption of households, to mobile tweets, to air quality data collected by sensors hanging in private homes' windows. We find geodata in the databases of map services like Open Street Map; there is an increasing number of Open Data published by municipalities and urban utility providers; civic sensor networks publish their data; most Tweets are geotagged. Also, there is hardly any technological limit to gathering and publishing geotagged data collected by sensors built into smartphones. The MIT Senseable City Lab is one of the first institutions creating awareness about the emerging areas of urban life covered by geodata and its potential uses through data visualisation and data analysis projects. One of their early projects mapping the density of mobile phone calls in urban areas created some iconic images.[4] Yet there still is a disconnect between understanding the bird's eye view visualisations of

---

[1]http://www.hellolamppost.co.uk/.

[2]Ingress by Ninatic Labs. https://www.ingress.com/.

[3]See Workshop *Instant Mapping in Rapid Game Design* by Invisible Playground. http://master.d esign.zhdk.ch/news/mapping-the-city.

[4]MIT Senseable City *Real Time Rome* project. http://senseable.mit.edu/realtimerome/.

geodata, and knowing how patterns of visualised data effect the street-level experience of urban life. This disconnect is reflected by the puzzlement of municipalities and mobility or utility providers who would be willing to publish data that is available to them anyway, but they seem rather unclear about what kinds of data in their possession is worth opening up as a meaningful contribution to the public. A better understanding of the research question addressed by the 'Size and shape of the playing field' (from now on Size and Shape) workshop, namely how geodata relates to street-level experience, should provide some clarification in this matter.

As the research through design approach generates knowledge through design activities, a more concrete design challenge will be introduced in this chapter, one that concretely embodies the research question. This design challenge pertains to the generation of the size and the shape of an ideal playing field for the urban game *Operation Noose* and will be described in more detail below.

As this chapter about the Size and Shape workshop series touches on three different nested levels of outcomes, these levels should be briefly clarified here and will be emphasised throughout the chapter. First, the main motivation behind the workshop and this chapter is to conduct and describe the experiment of employing game design in the *research through design* paradigm. Second, by nature, research through design has to address a *research question*: we have chosen the topic of how geodata can be related to the street-level urban experience. Third, as research through design is generating insight through design activities, we have chosen a concrete *design challenge* of the playing field generation that assumedly promises insight for the above research question.

2   From Design to Research and Back

The motivation to create the workshop format Size and Shape is a methodological one: the workshop format explores how design activities and reflection can generate implicit and explicit knowledge addressing a research question regarding a rather abstract matter. The research question about the relation of geodata and street-level experience mentioned above is the one addressed in the Size and Shape workshops, yet the aim is to contribute to a more general approach that can be employed for other research questions too. Advancing the concrete research question is a secondary goal, with advancing the *research through design* methodology by introducing game design being the primary goal. Nonetheless we still have to elaborate on the background of both the research question and the design challenge, to frame the Size and Shape workshop format. It is important to note that the design challenge was not originally created from scratch with the research question in mind; it emerged earlier in a different context and was chosen for this workshop based on the assumption that it would generate valuable insight for the research question.

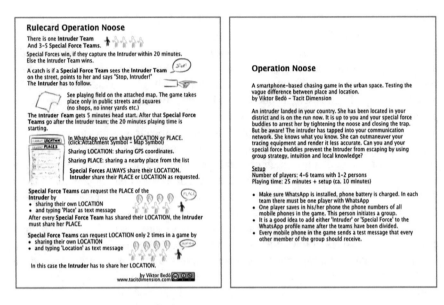

Fig. 1 Operation Noose rule card

2.1 The Design Challenge: Automated Playing Field Generation

The design challenge emerged out of a demand to allow players to autonomously start a session of the mobile-based urban game Operation Noose. Operation Noose is a chasing game where one player—*runner*—is on the run and is chased by three to five players—*chasers*. Players communicate with each other through an off-the-shelf messaging app like WhatsApp or Telegram. The app is used to share their position according to the rules of the game so that players are visible on the in-app map and to exchange short messages (Fig. 1). The game heavily relies on the principle of 'see and being seen' on the street as well as on the map in the messaging app. Although the map design and the way the position of the players is displayed on the map does have an influence on the gaming experience (see Coulton et al. 2017), Operation Noose relies on the in-built map of the messenger apps to minimise the overhead necessary for setting up a game session (Fig. 2). Every time chasers want to see the position of the runner, they all must post their own position first. The game's rules regulate when players have to share their exact GPS-based location or share vaguer position by choosing a nearby place from the list offered by the messenger app. Thus, the experience of the game is similar to the classic board game Scotland Yard, which plays with the accuracy and vagueness of knowing where Mr. X is at the moment. If one of the chasers manages to spot the runner on the street, the chaser makes an arrest by calling 'I have got you'. There is tagging involved as Operation Noose which is more about spatial tactics than about running.

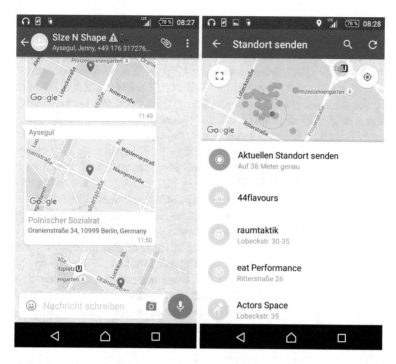

Fig. 2 Operation Noose messenger app screenshot during gameplay

All that is needed to play a round of Operation Noose is a couple of friends with WhatsApp installed on their smartphone, the rules, and a playing field of optimal size and shape. Yet the latter is a challenge: a too big of a playing field would make it too difficult or even impossible for the chasers to catch the runner, which would be unfair to the catcher. A too small playing field would result in catching the runner too quickly; this would be unfair to the runner and the game would be over too early. The size and the shape of the playing field is one of the main balancing factors for a fair and fun game experience.

The optimal size of the playing field is approximately 5 × 5 blocks, varying by give or take three blocks depending on the unique features of a given urban site. Such features, amongst others, are the length of corners, the curvature of streets, and the density of urban objects modifying visibility. Currently the playing field of any game session in a specific urban area is determined manually based on local knowledge. When we were running an Operation Noose session in Berlin, Zürich, or Malé for example, we had to rely on my personal experience as a game designer and knowledge of the cities to assign the optimal playing field. When Operation Noose was offered at events in cities where we could not be present and we did not know well, we had to rely on what we could read on maps of the unknown areas.

The question of an automated playing field generating emerged. An automated plotting of the playing field on a map would allow players to autonomously play the game anywhere using no further assistance than their smartphone. Ideally, the automatic playing field generator would make queries of local geodata and other relevant geo-databases, carry out an analysis of the data, then calculate and plot the size and shape of the playing field on a map for players to see. Wherever players are planning to play, they could use an app saying: 'This is where I stand, this is the number of players, show me the playing field around me (or a good one very close by)'. The challenge is to identify the kind of data that reflects relevant features of a given urban site, and which allow for automatically generating the optimal playing field.

Other urban games can pose the same playing field challenge. Any game would qualify that is played on the streets and where the size and the shape of the playing field (or playing time) are critical balancing factors for the game experience, and where the best choice of the playing area depends on the features of the given local urban environment. For example, in the game Horror Vacui[5] by Tacit Dimension, players move along the side walk on both sides of a street. One player uses his/her mobile phone camera to produce a continuous chain of photos about the other side of the street with no living person in the picture. Two counter players on the other side of the street do their best to appear on the pictures and thus invalidate the photos. In order to win, the player with the camera has approximately 10 min to get from the starting point to the finishing point and make his/her series of photos void of people, or else the counter players win the game. Depending on the frequency of non-playing pedestrians or the density of urban furniture that can blank out persons on the other side of the street, the playing time or the distance from starting point to finish has to be adopted.

In the workshop Size and Shape, the game of choice was Operation Noose due to its spatial scale and the grade of complexity of how urban features influence the gaming experience. Operation Noose covers a larger territory (app. 5 × 5 streets) than Horror Vacui (one or two blocks of a street). This design challenge exemplifies a context in which a better understanding of how geodata describes urban experience is highly relevant.

2.2 Research Question: How Does Geodata Link to Urban Experience?

Why should we better understand how geodata links to urban experience? Data is a key resource for driving the creation of innovative businesses and services that can deliver social and commercial value.[6] A better understanding of the potential implications of geodata on our urban lives can influence what is worth to investing in, opening up or building upon.

---

[5]http://tacitdimension.com/portfolio/horror-vacui/.

[6]See the work of the Open Knowledge Foundation. https://okfn.org/.

What does looking at geodata—or visualisations of geodata—reveal about what we can and cannot do on the streets, or in other words, what the street *affords* us? The term *affordance* was coined by Gibson (1979) and refers to how certain objects and setups in the environment implicitly reveal for attuned subjects the ways of potential interaction with them. Gibson provides a psychological account of the implicit interconnection of vision and movement when navigating the environment. The subject involved in a certain situation implicitly understands the potential ways of interacting with the environment, 'just by looking at it'. While walking in the street people mostly understand what it affords them when looking around. Walking toward the bus stop, a person, who is more or less aware of her walking speed, the distance to the bus stop, and the speed of the approaching bus, would be able to assess whether she can catch the bus. This person sees whether the situation affords the catching of the bus, or climbing the stairs, avoiding other pedestrians, etc.

To design for sites that we did not experience at the street level the question is: to what extent is it possible to infer the affordances of a site from looking at datasets in databases or urban geodata visualised on maps? What geodata allows us to infer to how it is to walk the street, to hide in the crowd, to visually control sections of the street? There is a disconnect between the street-level experience and looking at data available for the respective area. Clearly, looking at data or maps does not reveal the interaction possibilities of the respective portion of the street in such an immediate way as standing on the street and looking at the environment would. Yet as we will argue this disconnect can be bridged by an implicit understanding of how the two interrelate, and the question is how to deepen this implicit understanding. The disconnect actually touches on a classic cartographic problem of *how we understand maps*. Understanding how items and patterns that we see on the map relate to the mapped terrain is partly covered by what is called map reading, in the sense of knowing what map signs *represent* as highlighted by the legend, yet there is an aspect of understanding maps that is implicit (Robinson and Petchenik 1976, pp. 111–112). Elsewhere we argue that this implicit aspect of understanding maps that exceeds learning the explicit language of map signs is based on our fundamental ability to recognise similarity between what we see on maps and the embodied knowledge of urban life (Bedö 2011).

So we can argue that an expert would be capable of implicitly understanding the link between geodata and urban experience. How to become a better expert? What kind of specific urban experiences should an expert be exposed to in order to acquire the knowledge needed to relate geodata to the actual street experience? What activities and artefacts leverage implicit knowledge needed for this link? We argue that this is a key question to the research through design approach. To elaborate on this question, we have to have quick look at implicit knowledge, similarity and prototyping.

2.3 Implicit Knowledge and Research Through Design

The Size and Shape of The Playing Field workshop aims at leveraging and generating implicit knowledge to increase the expertise of workshop participants and enable richer reflection about the research question. Expert knowledge is generated both through the immersion of the participants in the urban environment through game-play and subsequent prototyping. The design activities' primary aim is generating insights related to the research question. This attitude is best described by the *research through design* approach, a subfield of *design research* in the design theory discourse. It is characterised by the creation of prototypes that instantiate hypotheses pertaining to a research question and by researchers and designers being involved together in the act of prototyping (Stappers et al. 2014).

Design research occupies the middle ground between basic research and design practice, for former dealing with more general knowledge within disciplinary boundaries whilst contextualization and multidisciplinarity increases towards the applied side of the range (Horváth 2007). On the applied side, the methodological rigour gives way to intuitive exploration through producing artefacts. As Stappers elaborates, research through design is the approach where artefacts and prototypes are produced with the aim being to generate knowledge about a research question:

> [T]he realization of working prototypes serves as the core axis connecting the design team or research team […]. In my view, this is the essence of 'research through design', i.e., that the designing act of creating prototypes is in itself a potential generator of knowledge (if only its insights do not 'disappear' into the prototype, but are fed back into the disciplinary and cross-disciplinary platforms that can fit these insights into the growth of theory[…]). But much of the value of prototypes as carriers of knowledge can be implicit or hidden. They embody solutions, but the problems they solve may not be recognized. There are two consequences of this: (i) efforts should be made to make these explicit, and feed this back to the informing disciplines; (ii) people working on related research topics should be exposed to the prototypes, so that they may learn solutions (gain associations) even if these are not made explicit. (Stappers 2007, p. 7)

We would like to complement this picture by emphasising that intuition—or, in other words, building on implicit knowledge—infuses all activities from scientific research to industrial design. While designers will naturally refer to intuition or implicit knowledge as the driver of their endeavours, science is widely associated exclusively with analytic methodology and formalised explicit arguments. The latter indeed characterises the description of research results once they are published. Yet implicit knowledge also has a decisive role in scientific knowledge generation (Polanyi 1966). A famous example is the discovery of the double helix structure of DNA by Watson and Crick. Besides their knowledge in chemistry and shadowy two-dimensional images produced by X-ray crystallography, the scientists hacked together a three-dimensional model of the molecule that in several iterations fostered the conciliation of chemical formulas with the architecture of the molecule (Hargittai and Hargittai 2000). Yet, as Gooding describes, the prototyping activity or even an image of the prototype did not get a prominent place in the publications of this discovery:

In science the role of images and artefacts changes, at first conveying a tentative understanding but later aiding communication and the construction of arguments. These generative and communicative roles usually fall away as new verbal and symbolic expressions become established. The declamatory caption given to the first published image of the double helical structure of DNA is a typical example. Although structural modelling had been essential to Watson and Crick's solution, their first paper does not include an image of the final model. Instead there is a diagram whose caption indicates that the image is purely an aid to understanding the written text. (Gooding 2004, p. 280)

Yet how do building models and prototypes support the reflection of more abstract matters? The key is that thinking and doing are different only in grade but not necessarily in nature and that we are implicitly able to establish the similarity of what we build and what we think. Schön's elaborations of design knowledge emphasise that there is reflection in action. To provide an example he describes the dialogue of an architectural professor and a student discussing a building that the student is planning (Schön 1983, pp. 79–93). The two look at the student's drawings, talk about them, change them while talking, looking at them again, and so forth. Schön says when the experienced architect looks at a sketch, he recognises similarities with design domains from the repertoire he acquired in the past, and an implicit iteration happens in a single moment: '[The architect] zeroes in immediately on fundamental schemes and decisions which quickly acquire the status of commitments. He compresses and perhaps masks the process by which designers learn from iterations of moves which lead them to reappreciate, reinvent, and redraw' (Schön 1983, p. 104).

The theory of tacit or implicit knowledge (Polanyi 1966) is often referred to as an explanatory theoretical framework when discussing intuition in the broader context of design. The reason for this is the wide acceptance of the fact that we acquire and use knowledge, which we cannot put into words or any explicit form. Tacit knowledge is embodied in our patterns of interaction with our environment and in our feel for stuff with which we are dealing (cf. Schön 1983, p. 49; Polanyi 1966). Theories of embodied knowledge provide an even more specific account of sensory-motor aspects of acquiring implicit knowledge through bodily interaction with the environment (cf. Noë 2004). To understand how interaction with the environment or with artefacts produces implicit knowledge and how this knowledge can become explicit and can be reflected on, we basically have to look at the nature of the contents of thinking: concepts.

In an empiricist approach to having concepts means the ability to keep experiences, being able to meaningfully reactivate them, and recognise similarities between actual experiences and past experiences (cf. Price 1953). Reactivating these experiences means the ability to produce mental and physical manifestations of the objects and events we have a concept about, so they can be the content of our thinking when the thing itself is not actually present. The philosopher Price emphasises that words are not the only instances that can be the content of thinking

Words and images are not the only non-instantive particulars which we use for thinking with, though in human beings they happen to be the most important ones. For example we also think sometimes by means of physical replicas, such as diagrams, models and dumb show; sometimes by non-imitative gestures, as in using the deaf and dumb alphabet; and sometimes by means of the muscular sensations which accompany incipient actions, gestures or others, when their actions are not overtly performed. […] In principle, any sort of non-instantiative particular might be used for thinking with, provided the thinker can learn to produce it for himself and to recognize it when produced by himself or by others. (Price 1953, p. 300)

So, in order to think with something one must be able to recognise and reproduce it. But thinking with something does not necessary mean having to reproduce it to a full degree. Sometimes it is not possible, sometimes it is not necessary to fully produce a manifestation of a concept, be it a word, a gesture or a movement.

Not only is the reactivation of a past experience a gradual affair, the reactivation may happen via words, movements, or even the readiness to move. Price calls the ability to reproduce something to even the smallest degree a *disposition*, which is a kind of readiness to produce an instance of our concept. Yet it does not necessarily mean that we are able to or that in the actual situation it is necessary to produce it on level of explicit words for example. Price describes how concepts are activated to various levels as a part of a network of concepts and features:

[A]s we have seen, the 'activating' of any mental disposition is a matter of degree. Between the two extremes – complete latency and complete actualization – there are many inter-mediate degrees of sub-activation. When the word 'cat' occurs, or a cat-like image, a whole series of concepts linked in one way or another with the concept Cat may be in some degree brought to mind. It is true of me at all times that I am capable of recognizing mice, bowls of milk, fur, tigers, mammals, hearth-rugs, at any rate so long as I retain a moderate standard of health and sanity. At all times I have memories of what all these diverse entities are like (in the dispositional sense of the word 'memory'). But if the word 'cat' occurs to my mind – or a cat image or a physical cat replica – then something comes to be true of me which is not true at all times. All these diverse memory-dispositions are in some degree excited or sub-activated. I am put into a state of readiness to recognize mice, bowls of milk, tigers, etc., if I should happen to perceive them; and also in a state of readiness to talk of such entities or produce images of them. I am ready to do these things, even though I do not actually do any of them. (Price 1953, pp. 317–318)

Things that are often enough experienced together will activate each other to the relevant degree when we recognise an item of this network. The better we know cats, the more details or other concepts are activated that are relevant to the actual situation. And these details might not only be mental images, but also patterns of inter-acting with the given object. So interacting with objects is obviously also getting to know them.

Noë argues in Perception in Action (2004) that touching objects as well as looking at objects is actually an action. According to his *enactive theory of perception*, recognising things by touch or vision is reenacting past active acts of perception. While we interact with our environment, internal sensorimotor patterns are produced based on our implicit understanding of how the perceived impulses from the environment interrelate with the proprioceptive understanding of our body states and body movements. It is these sensorimotor patterns that are reenacted when we recognise things or think about things. Noë also argues that these sensorimotor patterns are transmodal, meaning that on a certain level these patterns resemble regardless whether they were acquired through vision or touch. As a consequence, we are able to recognise things we see for the first time if we have touched the thing in the past. We argue elsewhere that transmodal recognition is also possible in the case when touch and vision are in different spatial scales: for example, when we recognise something on an urban map that we experienced in the streets (Bedö 2011).

We argue that the key to research through design is that in everyday life as well as in science and design we think in words as well as in actions and the production of artefacts. Adequate interaction with the environment, producing artefacts, and uttering meaningful words are all manifestations of knowledge and when we think and act these aspects activate and trigger each other like in a network as described by Price. Thus, finding the experiences to immerse in or creating the environments and artefacts to interact with is a vital part of reflecting upon more contextualised and rather abstract matters alike.

Also, the workshop series Size and Shape of The Playing Field explores a research question that benefits from the involvement of various fields of expertise, such as geodata experts and cartographers, urban designers and game designers. This exchange between various fields of expertise is again very much enabled by prototyping: 'Prototypes serve to instantiate hypotheses from contributing disciplines, and to communicate principles, facts and considerations between disciplines. They speak the language of experience, which unites us in the world' (Stappers 2007, p. 7).

3   Workshop Design

3.1   General Workshop Description

Size and Shape of The Playing field is conceived as a workshop series that will be run in various cities. The first invitation-only test run took place in Berlin on March 26, 2016; the first official workshop took place in Amsterdam in April 2016 as part of the workshop program of the Design and The City conference.[7] This subchapter provides an account of these two sessions, highlighting learnings from the events.

The core activity of the Size and Shape of The Playing Field workshop is determining the optimal playing field for the game Operation Noose in an urban area that the workshop participants do not know well. All other activities, such as playing the game Operation Noose for exploration and testing reasons and the reflection on the research question, are built around this core. The duration of the workshop session in Berlin was four hours and it was seven hours in Amsterdam. The workshop started with an introduction round in the workshop space to learn about the backgrounds and fields of expertise of the participants. The workshop host gave a short introduction about urban games, the aim of the workshop, and the principle of research through design in the context of the workshop. This was followed by two rounds of Operation Noose in the field. Back in the workshop space, the participants designed

---

[7]http://designandthecity.eu/programme/workshop/size-shape-of-the-playing-field-map-data-in-urban-game-design/.

a playing field for a new area with the facilitation of the workshop host, which constituted the core activity of the workshop. Having codesigned its size and the shape, participants went out again into the field to play Operation Noose and test the new playing field. The workshop was concluded with a reflection round in the workshop space about learnings and the experience.

In preparation for the workshop, two separate areas in the city were picked by the workshop host. In Berlin the playing areas were located in the Kreuzberg district, in Amsterdam they were located in the Wittenburg area and around Nieuwmarkt. The playing area was approximately the size of a neighbourhood, and the concrete borders of the playing field had to be defined within this area. In the first area a playing field was determined by the workshop host beforehand. In the second area the playing field was to be defined by the workshop participants in the design session.

In the workshops, right after the introduction, the participants' smartphones were set up: WhatsApp was installed on all devices that were to be used in the game and a WhatsApp group for the session was created. WhatsApp can be substituted with other messaging apps as long as they have the same location sharing functionalities, like Telegram. WhatsApp was chosen for the workshop as it was more likely installed on participants' phones already. The technical setup was followed by the introduction of the rules of Operation Noose and participants went into the field to play two rounds. Two rounds allowed for a learning curve in gameplay and thus the emergence of more elaborated tactical use of the game mechanics and the features of the urban environment. For most participants playing an urban game was a new experience, therefore the first round of Operation Noose felt like a trial round. Besides learning to master the game in general, in the first round players gained a better understanding of the affordances of the space, how fast they could move, and where they were exposed to other players. They also got a feel for how accurate positioning with their smartphones actually was, how many 'places' the map database contained around them. In the second round some of the experiences of the first round become more implicit as players gained proficiency in the gameplay so that they were able to focus more on the tactics and the exploitation of the urban environment. In the context of the workshop the game can be seen as a guided and intense exploration tool of a given neighbourhood. Through the high level of involvement in the game, participants quickly became experts not only in the game but also in respective features of the urban environment. Also, as players constantly looked at the map of the area they were playing in, they were sensitised for connecting the map view with the street-level experience.

The game session was followed by the playing field design session in the workshop space. The design challenge was to define the borders of a playing field in an area that participants had not been playing in before and which they did not know well. As participants did not have the street-level experience with the new target area, they had to rely on respective geodata available in the workshop space to determine the optimal size and the shape of the playing field.

It should be noted that, considering the aim of the workshop, participants ideally should not know the new target area at all: it should be a neighbourhood they have never lived in with a morphology and features they have not experienced before. For obvious travel time-related reasons the target playing area was relatively close to the first area, yet it was chosen so that its architecture and morphology provided a different experience than the first one. The main features that impact the experience in the concrete case of the game Operation Noose are related to interactions like hiding, assessing distances, and controlling an area visually. How far a player can spot another player, for example, depends on whether the game is played in grid like cities or continental city centres with curved streets, whether a crowd that is moving on the street, and the density of street furniture occupying the sidewalk. These affordances also might change occasionally, for example fields of vision being periodically obstructed by pedestrians who burst out of tube stations.

What participants at this stage, after having played Operation Noose, bring into the workshop is—besides their personal and professional experiences—the firsthand knowledge of what urban features matter to the gameplay. As the target area was different from the first area, the design of the playing field could just partly, but not entirely, rely on the game play experiences of the first round. During the playing field design session, the lack of experience with the target area was substituted with access to geodata made available in the workshop space, including suggestions from the workshop host. These suggestions were mainly taken from open data repositories. Workshop participants had a laptop to access data sources and some samples with short descriptions were also printed out on paper and made available in the workshop space. For Berlin the geodata suggestions included, amongst others, the position of elevators in public transport facilities, the position of glass recycling containers and the position of all trees.[8] For Amsterdam the suggested information included a map that showed the detailed shape of buildings and a map showing areas where pedicabs are allowed or prohibited.[9] The connection between geodata and street-level experience was approached from two angles. First, participants collected relevant features of the urban environment and checked whether they could find data or maps that would allow inferences about these features. Second, coming from the other direction, participants looked at examples of data and maps to see whether they resonated with any of the experiences they had in the field.

The negotiation of the relevant features of the target area and their effect on the ideal playing field are supported by a printed map of the target area (scale app. 1:5000) and some materials (dots, pins, lego, etc.) to mark and highlight data, interactions or features of the area on the map (Fig. 3).

Using these materials, the experience from the game session, and their expertise workshop, participants negotiated the size and the shape of the playing field for the second session. The paper map allowed the use of pins and a thread spanned between them to flexibly mark and manipulate the shape of the playing field on the map.

---

[8]http://daten.berlin.de/datensaetze.

[9]http://maps.amsterdam.nl/.

Fig. 3 Prototyping desk Berlin

3.2 Berlin Test Run Session

One of the insights from the first two rounds of Operation Noose was that *accessibility* is much more relevant than the categories of public or private space. For the players the pertinent question was whether it was physically possible to walk into an inner yard of a housing block, regardless of whether the yard was administratively public or private. Hidden paths, artificial or natural elevations, dead ends, and temporal building sites were urban elements and features that were identified among others as relevant to gameplay.

Based on these findings, participants conducted a very short desktop research session to identify data that allowed an assessment about passages through blocks and foot paths between building blocks or within areas enclosed by building blocks. Based on the playing experiences and the available data participants determined dimensions of the new playing field in the target area (Fig. 4).

To test the assumptions made in the design session, participants went out into the field to play another round of Operation Noose in the new playing field. This was followed by a reflection round about how far the test round validated the assumptions of the design session. Following the design session, the subsequent game affirmed the relevance of foot paths, as the playing area included several residential housing blocks from the 1950s with strongly compartmentalised semi-closed inner yards. Small playgrounds, elevated flower beds and sheds for rubbish bins constituted a landscape architecture that did not reveal itself by looking at these blocks on the map in the prior design session. The morphology of the landscape architecture within the morphology of the blocks created a playing experience that was described by participants as a quasi-fractal structure with smaller compartments

Fig. 4 Playing field prototype Berlin

in the bigger blocks. Such an urban structure would motivate the choice of a smaller playing field as the level of spatial complexity that the game requires is provided by a smaller area.

Fences were revealed as a key factor influencing possible shortcuts and detours during gameplay. The mesh wire fences did not obscure vision drastically, yet they did prevent players from crossing an area. Participants noted that fences were rarely mapped as they were installed and maintained by the fragmented owner structure of the buildings. Trees might be relevant if they are in a row, as players cannot hide behind a single tree, yet participants agreed that a row of trees alongside the street or a promenade did change the conditions of visibility and increased the chances for the player on the run to escape. So the density of trees in general would be less of a relevant factor, than rows of trees. As such, the presence of rows of trees in the field might require a smaller playing field so the chances for the chasers to spot the player on the run are balanced out. Participants also agreed that lower but denser vegetation, like bushes, had a much larger effect on the visibility and the chances of being spotted than trees. The Berlin open data portal does not have data about bushes—whether this kind of information is available for Berlin from another source remains to be seen. Some factors had a significant temporal character. Crossing larger streets used a lot of time in the field, as players had to wait for the green light or the possibility to cross when no cars were approaching. As far as participants were aware, no open data existed about the frequency of the traffic lights at the pedestrian crossing. Other seasonal features discussed included the queue in front of an ice cream shop.

3.3 Amsterdam Session

Based on the first two rounds of Operation Noose in the Amsterdam session, the following features of the playing area were identified as relevant: parking lots are good places to hide; maps show whether the facade of a building is flat or it has corners; and structures that allow for hopping in and out when hiding. Old Amsterdam buildings often have a couple of stairs leading up to the front door and also have stairs leading to basement premises. Bus stops and trees were discussed as objects obscuring vision, making it easier to hide. Most prominently, a semi-public inner yard system of a residential housing block in the initial playing field increased the 'fractal structure' of the area. Not accounting for these inner yards and only counting the number of housing blocks on the map would clearly result in a less ideal determination of the playing field size.

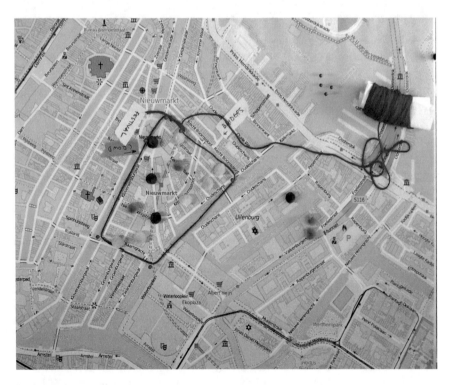

Fig. 5 Playing field prototype Amsterdam

Fig. 6 Operation Noose session Amsterdam

After sharing reflections about the initial playing, field participants turned to the design of the playing field for the target area. As the target area—as opposed to the initial area—was located in one of the most central and popular areas of Amsterdam, the crowdedness of the street very quickly dominated the design process of the size and the shape of the playing field. This was also almost the only feature—next to channels and the frequency of bridges—that was marked on the map. With a colour coding, the streets were ranked in tree levels of crowdedness. The density of shops and metro stations was discussed as the main indicator for crowdedness.

In the workshop space, the assumed indicators for crowdedness were the density of shops in certain streets, metro stations, and a festival that was taking place in one corner of the target area. Based on these factors and their assumed effect on the gameplay, participants negotiated the borders of the playing field (Fig. 5) and went out to play (Fig. 6).

4   Learnings and Implications

The design challenge of generating the playing field has probably seen the most immediate outcome of the two workshop sessions. The insights generated in the workshops could easily feed into design patterns for automatic playing field generation for

Operation Noose and games of similar format. This is not a surprise as the codesign session of the playing field was the focal point of the workshop and workshop participants had an easy time engaging with the session as well as with the gameplay.

The workshops provided some insights about the research question, namely how geodata relates to the street-level experience. One of the most explicit discussions about the relevance of open data for playing field generation was the position of trees in the streets of Berlin: participants agreed that while a single tree does not affect the gaming experience much, a line of trees can already shield vision from certain angles and thus make it easier to hide. In the concrete case of the playing field generation this would mean that the optimal playing field has to be smaller. More generally participants developed an understanding of how the spatial layout of trees or urban furniture of a similar form influences the visual permeability of a street or a square. Much of other open geodata that was considered in the workshop, like the location of recycling containers or the location of elevators at train stations, can be useful to help citizens find these facilities. However, participants did not find them to be useful references for the affordances of a certain urban site as described above.

In negotiating the size and the shape of the playing field in the codesign part, the data available in the workshop space clearly played a secondary role compared to the experience participants gathered in the field. The abstract data could by far not compete with the immersive experience during gameplay, towards which the argumentation in the codesigning process gravitated. Participants with more local knowledge were clearly relying on their everyday knowledge of the area so that the threshold for considering the exploration of datasets was rather high. It proved to be useful to explore the relationship between geodata and the street-level experience from both sides: first looking at the data to ask ourselves what it can say about relevant features of the urban environment and then looking at relevant features and asking what data can tell us about it. Both perspectives need to be adopted and the design session should explicitly allocate time for both perspectives. Future workshops could reveal how the weight of geodata in the codesign process might change with workshop setups that allow for a more elaborate and deeper involvement with the data and more participants with backgrounds in GIS and cartography. As the research workshop will be run in different cities, it also provides a comparison between the density and quality of accessible data in these cities.

Workshop materials in the codesigning session were key assets to permeate the border between the implicit knowledge of the participants and the explicit discussion. The map, and various objects that allow marking of the map, were tools for making hints and cues tangible to other participants, not only to communicate but also to think together. These materials allowed for both datasets to be visualised and the features of the urban space as experienced by individual participants and created a kind of model that made it easier to bridge and negotiate the street-level experience with the raw lists of geodata. Nonetheless, the gap between the data and the street-level experience or affordance of the urban sites remains. Also, according to participants' accounts, the gameplay was so engaging that they hardly ever explicitly focused on those features of the urban environment that were discussed in the workshop space. It is an obvious challenge to objectively distinguish which arguments in the discussions were motivated by which implicit aspects of the participants' backgrounds and immediate experiences with the game. Future workshop might be extended with qualitative methods to capture and analyse this matter in more depth.

After the game sessions and during the negotiations of the size and shape of the playing field participants often suggested changes in the ruleset of the game, some minor changes were implemented in subsequent rounds. These circumstances support the claim that game design-based challenges can optimally be employed in the research through design approach, as games can be easily adjusted according to the actual research question and grant high engagement with respective features of the urban environment.

5   Conclusions

A better understanding about what geodata tells us about the urban experience can drive the debate about what kinds of geodata should be published openly as a meaningful contribution to the public. Following the research through design approach the Size and Shape of the Playing Field workshop series uses an urban game design-related challenge to explore how we can bridge geodata with urban experience. For the workshop participants an urban game provides a genuine and engaging immersion into aspects of the urban environment that can be defined by the game mechanics. Thus complementary to the knowledge based on participants' personal backgrounds, the gameplay builds a shared expert knowledge that and that can be built upon in the codesigning activities in the workshop. Workshop materials and joint prototyping aim at bridging the playing experience with the geodata by fostering implicit manipulation and explicit negotiation about the optimal size of playing field prototype.

Literature

Bedö, V.: Street game design: rapid prototyping for the urban. In: de Waal, M., de Lange, M. (eds.) Hackable Cities: From Subversive City Making to Systemic Change (Forthcoming 2017)

Bedö, V. Interaktive Stadtkarten als Instrumente der Erkenntnis. PhD Thesis, Humboldt-University Berlin, http://edoc.hu-berlin.de/docviews/abstract.php?id=38969 (2011). Accessed 25 June 2016

Coulton, P., Huck, J., Gradinar, A. et al.: Mapping the beach beneath the street: digital cartography for the playable city. In: Nijholt, A. (ed.) Playable Cities. Series: Gaming Media and Social Effects. Springer, Singapore (2017)

Gibson, J.J.: The ecological approach to visual perception. Houghton Mifflin, Boston (1979)

Gooding, D.C.: Envisioning explanations—the art in science. Interdisc. Sci. Rev. **29**(3), 278–294 (2004)

Hargittai, I., Hargittai, M.: In Our Own Image: Personal Symmetry in Discovery. Kluwer, New York (2000)

Horváth, I.: Comparison of three methodological approaches of design research. In: Proceedings of International Conference on Engineering Design, ICED '07, 28–31 Aug 2007 (2007)

Nijholt, A. (ed.): More Playful User Interfaces. Interfaces that Invite Social and Physical Interaction. Series: Gaming media and social effects. Springer, Singapore (2015)

Noë, A.: Action in Perception. MIT Press, Cambridge, MA (2004)

Polanyi, M.: The Tacit Dimension. Routledge, Kegan Paul, London (1966)

Price, H.H.: Thinking and Experience. Hutchinsons's Universal Library, London (1953)

Robinson, A., Petchenik, B.: The Nature of Maps. University of Chicago Press, Chicago, London (1976)

Schön, D.: The Reflective Practitioner: How Professionals Think in Action. Basic Books, New York (1983)

Stappers, P.J.: Doing design as a part of doing research. In: Michel, R. (ed.) Design Research Now: Essays and Selected Projects, pp. 81–91. Birkhauser, Basel (2007)

Stappers, P.J., Sleeswijk Visser, F., Keller, A.I.: The role of prototypes and frameworks for structuring explorations by research through design. In: Rodgers, P., Yee, J. (eds.) The Routledge Companion to Design Research, pp. 163–174. Taylor & Francis, Oxford (2014)

Game Engines for Urban Exploration: Bridging Science Narrative for Broader Participants

Verina Cristie and Matthias Berger

Abstract One aspect of playing is exploration. A playable city could therefore be regarded as an explorable city. In recent years, a growing number of urban exploration tools have been developed, empowered by the technology of game engines. Traditionally, game engines have been used to create virtual environments for entertainment and enable the user to explore. We seek to apply game engines in urban planning beyond the visualization of buildings, trees, traffic, or people in the city. It can become a tool for a multidisciplinary approach that involves engineers, scientists, architects, planners, and even the citizens themselves. Those kind of mixed stakeholders have quite diverse needs, which all can be addressed by game engines in an easy way. In this chapter, we look toward bridging the seen and unseen elements in a collaborative game environment. One of the less visible elements in urban environment involves the urban microclimate: heat emission, wind flows, and outdoor thermal comfort. These unseen scientific elements have become a narrative on their own on top of the urban exploration. If they are to be presented in the traditional way of scientific visualization, the connection to the built environment would be difficult to understand for many kinds of stakeholders, especially the nonscientists. Hence, we utilize the power of the Unity3D game engine to show the potential of collaborative and explorative virtual environments: a bottom-up citizen design science within the narrative of exploration and urban science data.

V. Cristie (✉)
Singapore University of Technology and Design SUTD, 8 Somapah Road,
Singapore 487372, Singapore
e-mail: verina_cristie@mymail.sutd.edu.sg

M. Berger
Department of Architecture, ETH Zurich, Building HIT, Wolfgang-Pauli-Str. 27,
8093 Zurich, Switzerland
e-mail: mberger@arch.ethz.ch

M. Berger
Future Cities Laboratory, Singapore-ETH Centre, 1 Create Way, #06-01 CREATE Tower,
Singapore 138602, Singapore

© Springer Science+Business Media Singapore 2017
A. Nijholt (ed.), *Playable Cities*, Gaming Media and Social Effects,
DOI 10.1007/978-981-10-1962-3_5

87

Keywords Citizen design science · Game engine · Unity3D · Urban microclimate · Urban planning · Visualization

1 Introduction

What is a playable city? By starting on what it is not one can determine which elements are conditional, which elements are enabling, and where the barriers are. In terms of urban planning a top-down planning approach might exclude citizens' participation and knowledge. It transforms citizens from agents or protagonists into patients, sufferers of their fate. 'Playable' incorporates a bottom-up approach of citizens engaging in an active, positive loop with their city. While we assume for citizens that a certain amount of interest and willingness exists to interact with [the] urban planning [authorities], there is a need to bridge the gap of data, information, and knowledge required to actually do so (Berger 2014). Furthermore, channels of communication between the stakeholders are essential to promote a functional discourse, and skills to use them require training as well. The traditional top-down approach in urban planning is being replaced by the new forms of mobile and cooperative labor.[1] Following Pierre Levy, it is important to have capabilities of collective intelligence (Lévy 1997); to compare, regulate, communicate, and reorganize one's activity. Our vision is that game engines can provide the key to enter the world of urban planning for citizens while reducing the need for previous knowledge and skills at the same time. With game engines citizens can even contribute to citizen design science,[2] give feedback to urban design and planning directly (Crowston and Prestopnik 2013).

Four out of five households in the U.S. own a device to play video games, reports *Essential Facts About Computer and Video Games Industry* published by the *Entertainment Software Association* in 2015 (ESA 2015). The big market for computer games together with Moore's law explains the constant need for progress in game engines: state of the art computer graphics namely the lighting, texturing, particle systems, physics, animation, and special effects, all blended in eye gratifying visuals.[3] As such, we can see game engine as a very powerful visualization tool for nongaming content as well. Furthermore, game engines are commonly made for license-free end-usage (even so programming requires a proprietary license) and are platform independent. The Unity3D engine, i.e., runs on Windows, OS X, Xbox 360, Xbox One, Wii U, New 3DS, PlayStation 3, PlayStation 4, PlayStation Vita, Windows Phone, iOS, Android, BlackBerry 10, Tizen,

---

[1]Paul Coulton (University of Lancaster, UK) notes that new policies in the UK require now all towns and cities to include participation.

[2]For a working definition of citizen design science see Wikipedia (2016).

[3]Countless examples could be given here. The work of Ulrike Wissen (Manyoky et al. 2014) is most comparable: simulating audio-visual landscapes with the CryEngine.

Unity Web Player, Windows Store, WebGL, Oculus Rift, Gear VR, Android TV, and Samsung Smart TV. Humanity is moving toward having more and more *digital natives*, a term connected to the name Marc Prensky; his team bridged a new CAD software learning experience by creating a game like tutorial (Prensky 2001).

In a similar manner, we believe that it is feasible to use game and play like behavior to bring urban science to a wider audience. This chapter will expand the role of game engine not only as a tool for urban exploration, but also as a tool to tell a narrative within the context of urban science.

1.1 Games as Narratives

Game design is often seen as an interplay between interactivity and narratives (Ampatzidou and Molenda 2014). However, not all games are telling stories, although many games have narrative aspirations (Meirelles 2013). We need to expand the concept of games beyond its definition as a system to win or following objectives, given that different elements interacting with each other in somehow limiting constraints (Abt 1987). Helpful might be the differentiation made by LeBlanc, which separates emergent narratives from embedded narratives (LeBlanc 2000). Emergent narratives are defined as storytelling that comes as a result of player actions and events within the game. The embedded narrative is the [background] storyline, content which has been pre-generated prior to the player's interaction with the game.

The creation of embedded narratives includes the creation of hypothetical, fictional worlds [or realms, levels], so a big portion of game design is level design, where worlds are designed and spaces are sculpted. If the context of the fictional world is a city, then the level design becomes similar to urban design, yet a simpler one—without having to consider the most realistic representation of buildings and their layout within the city. Since we want citizens to become players [who are supposed to walk through the spatial environments of the virtual representation of urban topologies], we need to look into emergent narratives of the urban planning environment and compare them with game narratives.

1.2 Citizen, Urban Design, and Science as a Narrative

The Chair of Information Architecture at the Swiss Federal Institute of Technology in Zurich (ETHZ) is dealing exactly with the interplay of *Citizen, Urban Design, and Science*, since it has been established around 2006: "We develop visual methods for the analysis, design and simulation of urban systems for a sustainable future" is the leitmotif of the research group. Citizens as well as urban planning authorities are equally seen as stakeholders. By the nature of the specific problem, often one stakeholder has or should have more power, i.e., for an airport superior interests of a larger region predominate and residents' NIMBY (not in my backyard) behavior would rather destructive than constructive, whereas the design of a local playground explicitly should consult the residents. While the discourse or dialog between citizens and authorities has to be direct, we can provide tools and means to enable, enhance, and democratize—by given equal access to data, information, and knowledge—the communication.

Here computer-aided tools come into play. Exploration tools allow users to play and incorporate naturally the concepts of [video or computer] games. Perspectives as first-person or birds-eye come into action as players walk through different places of the virtual environment (Zeng et al. 2014). Collaborative tools come in form of multiplayer games, or even as a person observing other people playing a game and voice out his/her opinion (Klein et al. 2014). Communication from virtual to real world is a feature here (Treyer et al. 2013). Not only the exchange of information and knowledge has to be taken into account, the requirements for the exchange of data from real world and scientific domain has to be included during design and implementation of a tool.

In this context, the data then becomes the narrative in or of the urban landscape. Will more trees in the city provide a better climate such that residents are more satisfied? How does putting a hospital at the city center correlates to the longer life expectancy in the city? What will happen if factories are being relocated to the other part of the city? As such, simulation tools in the urban modeling have provided a dynamic context to urban narratives (Guhathakurta 2002). Not only that it has created a method to construct 'stories' of the past and possible futures, but it is also able to create stories of the current time with multiple [design] alternatives.

2 Related Works

2.1 Game Engines for Urban Explorations

In *visualisation support for exploring urban space and place* (Pettit et al. 2012) a game engine is being categorized as visualization support tool together with GIS and cartography tools, digital globe (Google Earth, ArcGIS explorer), virtual simulation environments, and building information models (BIM). The development of game engines as urban visualization tools comes from the need to have a user perspective as well as navigational view. These views are understood to be more engaging as they contain more complex spatial information to be explored and learned (Clemenson and Stark 2015). Other than the purpose of engagement, having a 3D navigational view could also help user from missing out information that could have been understood or caught only if 3D navigational view was available. Such examples include the designing of rural roads in which many curves are needed in uneven topography and heights. By navigating through the road from the level of different vehicle's height, for example, critical blind sections that would cause accidents otherwise can be identified and minimized (Kühn and Jha 2006).

We noted that there are various game engines available in the market, of which better known to urban exploration are Unreal Engine and Unity3D. Both have been shown in many cases of serious gaming, as in various training and educational fields. Several applications have been made related to virtual urban environment navigation using Unity3D, such as Yaesu district, Tokyo (Indraprastha and Shinozaki 2008) and the 19th century town scale model of the Lorraine region, Paris (Humbert et al. 2011). Similarly, Boeykens also coupled the tool with 3D design software and a BIM to do historical reconstruction and to finally explore the environment (Boeykens 2011). Game engines provide additional benefits as adding a 'wow' factor in BIM presentations and visualizations (Slowey 2015). This trend in the industry is also boosted with the current launch of Autodesk's Stingray game engine. In additional, not only visual, but also sound representations in the public space could be explored and evaluated in the work done by Signorelli (2011).

The Unreal Engine has been used for urban scale visualization in the Paestum tourism area, Italy (Andreoli et al. 2006). In this work, the real position of user is to be reflected in the virtual environment by using location awareness technology through GPS and Wifi. The engine is also capable of visualizing indoor and outdoor elements (Fritsch and Kada 2004). Both engines have a powerful rendering pipeline as they are built for games; hence, an interactive frame rate is achievable. They could also run on multiple platforms of operating systems, and could be exported to mobile devices. By nature, 3D models are easily imported. Lighting system is available to provide vivid scenery. Special effects can be added to enhance visual esthetics of the visualization. For outdoor element, terrain and tree systems are available in abundance. Networking frameworks are built in the system, as the engines are able to support multiplayer or network gaming. In addition, basic physics engine that deal with gravitation, Newton's laws, particle systems, even fluid dynamics are available such that a scientific simulation could be performed. In Xu et al. (2013) computed a dynamic fire simulation to show how smoke propagates in high-rise building in case of fire. They built upon the existing game engine's particle system and improved it by adding a stack effect model and computing the coordinate system algorithm using Excel file. While both the engines are equally powerful, in our study cases, we used Unity3D as it has lower learning curve for our programmers who are more familiar with Unity3D with C# than Unreal with C++ (Unreal uses C++).

2.2 Scientific Visualization Tools

Some background is needed to emphasize the difference between scientific visualizations and [general] information visualizations. Information visualization in cities is usually related to abstract data representing the elements of a city to understand the larger concept of a city. Such data includes for example population data, infrastructure data, and economic data. The data that we will be looking at in our case studies is related to the [urban] science of climatology, such as heat [in temperature and humidity] and wind flow [as velocity and direction], where the geographic location is given [in our case study it will be Singapore] and simulations of the local microclimate are performed (Papadopoulou et al. 2015). In the most straightforward visualization manner, the process could start from data collection, continued with the 3D modeling of the data, and presentation (Kim and Bejleri 2005). Challenges are everywhere, as data collection deals with an accuracy issue, 3D modeling deals with a reality issue, and visualization with its representativeness (Wissen Hayek 2009).

We experienced that especially in the architectural world; there is a tendency for scientific visualizations to be separated from the design process. Visualizations are conventionally seen rather as the end product than as a part of designing itself, e.g., conducting wind flow studies for the purpose of obtaining scientific-looking evidence[4] rather than optimizing the design subject to wind.

2.3 Public Participatory Urban Planning Games

Participatory design is meant for an inclusive planning with all stakeholders. In this process, they are able to negotiate, to reach an agreement, allowing flexibility and the learning curve to progress over time for stakeholders of different backgrounds (Chen and Schnabel 2011). In a similar note, collaborative design is viewed as social action where instrumental, communicative, discursive, and strategic actions take place. Instrumental action includes drawing and producing prototype. Goal orientation is considered discursive, and negotiation is considered strategic (Vosinakis et al. 2008).

There have been several discussions on bringing the city planning exercise into game like environments where different stakeholders get to negotiate to reach a consensus. This could be done either in a physical or in a virtual space. While it is traditionally easier to communicate in physical space, there are so many avenues to communicate in digital space nowadays. One such digital/virtual example is *YouPlaceIt!* (Vemuri et al. 2014). It brings a serious game play approach and has to be differentiated from other city planning games whose purpose is solely for entertainment. In *YouPlaceIt!* each player will take a role in urban planning [like resident representatives, government, real estate agents, NGO, etc.], and discuss through multilingual text chats and icons about an objective in a study case in Dharavi in Mumbai, India. For each action done, the game engine will calculate the costs, all in all leading to a consensus. While we advise to relay on participatory processes in general, the timing is even more crucial: Including end user early on in the design process empowers richer design solutions (Bodker et al. 2004).

In *Nextcampus*, early public engagement through an online serious game was developed to find a new location for current university campus. Each player is a resident of the campus and he/she is given an initial amount of money to further do

---

[4]The Bahrain World Trade Center is here an example where environmental ambitions became a selling point, with three wind turbines integrated into the building, even so wind flow studied would have led to a different result: the turbines are facing the wrong direction.

the planning and observe the mood of the local population [= the campus' residents]. The goal is to find the best strategy of relocating that will ensure a good mood in population (Poplin 2012).

The playful element through game is deemed necessary as a potential solution because the public are in general rationally ignorant (Krek 2005) due to the high cost [e.g., time and difficulty] to be involved in the planning and also to learn about planning alternatives. Hence, through serious game, participant could be engaged in learning and participate in the planning unintendedly. Some other examples of serious game would also be usually deployed in the city museums where more audience could be reached. One of the examples is SingCity. It is a game installed at Singapore's city gallery that will invite up to eight participants to solve the needs of the residents in city planning (URA 2016).

Adapted from (Cecchini and Rizzi 2001), we then can compile the following list of requirements for an urban 'gaming' simulation for participatory urban planning, in contrary to pure video games with urban settings like the notorious Grand Theft Auto (GTA):

- Made for (and with feedback by) lay persons which are as well end-users
- Standardized interface to other software
- Providing control to the user, enable scenario-making and evaluation
- Not include 'give' knowledge or results, user has to come to own conclusions
- Interactivity to be based on suitable menus (software) and human–machine interfaces (hardware)
- No proprietary hardware required, better able to run on any device
- License-free, freeware, shareware, or General Public License (GNU).

2.4 Virtual Reality and Interaction Tools

The concept of virtual reality in architecture came in the early 1990s as proposed by Schmitt et al. (1995). Since then, exploration of space has been advancing in the virtual world using computers without the need of a physical 3D model representation or 2D drawing based on pen and paper.[5] Research on how to interact with the virtual world has also been progressing from Graphical User Interfaces (GUIs) to natural user interfaces. Tangible interfaces are now expected to bridge the gap between the real world and the virtual world. As such, an augmented reality on tangible interfaces could be used rather than a fully virtual reality (Müller Arisona et al. 2012). In urban design collaborative platforms, i.e., a pen-like 3D manipulator or direct manipulation of 3D blocks could be employed to modify the urban layout, which is then projected to a head-mounted display or display screen (Seichter 2007).

---

[5]This is one direction, coming from the real world into the digital domain. Vice versa, with augmented reality (AR) both domains could be merged with the physical domain in focus, e.g., https://vimeo.com/65130490.

'Play the City' uses a games physical environment as method for collaborative decision making by bottom-up stakeholders (Tan 2015). By having augmented environment, another dimension of city planning can be added: the unseen elements of traffic simulations or climate related studies. Such example includes 'Tangible 3D Urban Simulation' table by RMIT university to simulate urban wind using projection and fabricated urban 3D blocks (Salim 2014), or 'CityScope' by MIT Media Lab to simulate urban metrics like walkability, energy, daylighting, and trip upon changing the neighborhood block configurations using projection and Lego® blocks (Winder 2014).

Fig. 1 The context of the environment is shown with a multiscreen setup. Users are able to focus on different aspects of the environment and adjust the views according to what they need. The views on the screen build a cohesive understanding of a story the presenter wants to portray

Table 1 Interface and collaborative space options in physical, augmented, and virtual environments in architecture and urban planning

	Environment		
	Physical	Augmented	Virtual
Interface	Tangible interface: 3D model, pen, and paper	Tangible interface with head-mounted display/projectors (usually table-top-based)	Any display device or screen
Collaborative space	Same room		Could be same room (using a large screen) or via internet (multiple players, each with own screen)

For a fully immersive experience, large displays or projectors in a CAVE-like environment are still preferred for urban planning or visualizing project models. In addition to a big amount of information that can be displayed on the screen, such setting allows multiple participants to discuss and collaborate with each other over the screen's content. ETH's *ValueLab* exemplified an environment for collaborative environment: Five large screens with a total of 16 mega pixels and equipped with touch interface capabilities and three full-HD projectors as visualization and decision support tools in urban planning (Kunze et al. 2012). Figure 1 shows the latest multi-screen setup in the *ValueLab Asia*, where the projectors have been replaced by multi-touch screens. Alternatively, a head-mounted display is utilized for augmented reality projections. The cons, however, are that they are heavy and prolonged usage can cause discomfort. In the spirit of play and exploration, one also can look forward to the recent advancement of omnidirectional treadmill as developed by *Virtuix* and *Cyberith*. Just imagine roaming around in the virtual city with walking and even jumping. Another advantage with a fully immersive virtual environment is also that collaborative action does not need to happen in the same physical space. Game engines, especially, have the capability to provide Massive Multiplayer Online Role-Playing Game (MMORPG) experience. Some of famous multiplayer online city building game are *SimCity* and Sid Meier's *Civilization*. Table 1 summarizes the classification we use for the different kinds of environments according to interface and setup.

3   Data Driven Design Interfaces

3.1   Operations with Data

With the availability of open and big [urban] data through online Application Program Interfaces (APIs), various visualization tools have been developed for a collaborative urban planning of future. One of such is *OSCity* (van der Net 2013) in which we note its two key features: 'Urban Dash' to monitor spatial events and trends, and 'Spatial Search' to have semantic search within GIS analysis, both in real time. A semantic search capability is useful to transform natural language to GIS queries. As such, for example, one will be able to search supermarkets with certain names, or offices along the river. Major challenges in big data visualization are perceptual and interactive scalability (Liu et al. 2013). The number of data points visualized must be balanced such that it will not overwhelm user's perceptual and cognitive capabilities. This balancing is also important because we do not miss any interesting structure or outliers. The amount of data loaded furthermore will determine the latency of the tool, finally affecting the interactivity of the exploration.[6]

---

[6]The MATSim framework (http://www.matsim.org), for example is an excellent transport planning tool, but has never been designed for an interactive use in workshops. The whole simulation has been developed with the underlying assumption that the solver can take more time, yielding a better accuracy, and computing more complex and large cases.

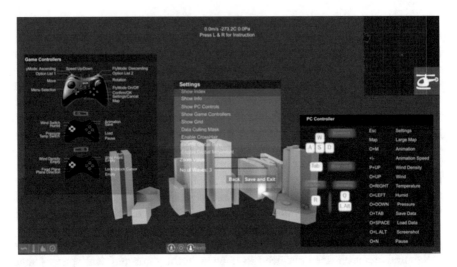

Fig. 2 There could be more than one set of navigation and control for 3D exploration. Here, keyboard controls and a joystick are allowed. For navigation, the mode is set to 'helicopter'—a virtual fly-through. The map view is located at the *top right* corner to easily find the current position

There are several guidelines in creating a user interface for an interactive environment exploration. It is important for user to be able to navigate around not only by an arbitrary or third-person camera view but also by walking on pedestrian level to simulate realism. Next, the user should be able to plan the views. For example, the user should be allowed to save the camera's position and create a fly-through view as in Fig. 2. Views can be saved as screenshots, and fly-through can be saved as video. This view will especially be helpful for presentation and discussion. Knowing current coordinates by using a map will also help user to navigate better: to put emphasis on the data [about the urban microclimate] in our tool, buildings are visualized as grayish, featureless blocks (as in Fig. 5); recognition of real world locations is hence limited. As such, like in first-person games a 2D map view can complement the 3D view. We also noted several optional interactions with the 3D model. Such as mentioned by Greenwood et al. are: importing the 3D model, transformation of model (scale, move, rotate), measurement of model, copy and delete, layering of models (to better organize the visibility of model), and editing of model's texture. In additional, the tool allows the user to analyze shadows and line of sight in the model (Greenwood et al. 2009). In Fig. 3 the menu for customization in our tool is shown.

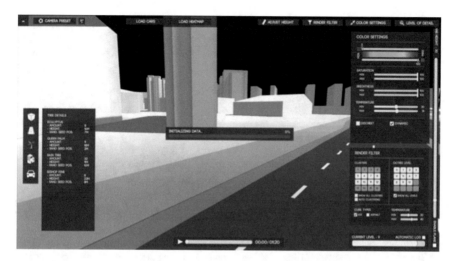

Fig. 3 Camera presets and views are necessary elements in every exploration tool. The info panel helps users to know the context of data and model. Filter and color settings help further to understand the information shown better

3.2 Realism, Art, and Style

Architectural design is a visually evaluated process. Onur Yüce Gün stresses the importance of realism in the architectural process in his work on a very real walk-through using interactive real-time rendered stereoscopic animations (Gün 2013). Realism is described as the resemblance of the displayed image versus the reality. The degree of realism depends on its level of abstraction. Abstraction levels can go from very similar (reality) representation, to indexed, iconic, symbolic, and down to language writes Lars Bodum in Dykes et al. (2005). In Zanola et al. (2009), the authors created three different levels of realism in a 3D stereoscopic display of a spatial data, namely photorealistic, CAD-style, and sketch-style rendering. After the experiment where users were shown the representation on a video wall, they were asked of their confidence of the credibility of the spatial display. Photorealistic has higher confidence level than CAD-style, and CAD-style has higher confidence level to the sketch-style. Clearly the higher level the realism is, the higher the confidence level will be (Manyoky et al. 2014) (Fig. 4).

It is common understanding that the higher the realism is, the less interactive the tool will be. For example, a realistic model will result in bigger file size of geometry, which then can make the rendering slower. Game engines, however, deal with rendering quite well since an interactive frame rate is a must have in every game. It does this by optimized utilization of the GPU to process the rendering faster. Another level of detail and realism, adding vegetation is also understood to improve the quality of the environment and creates an esthetically more pleasant picture. Depending of the art and style that we aim for, we might

Fig. 4 Public transport data is visualized as animated *green lines* over nontextured building models (© Treyer 2014). This mock-up was highly acknowledged by visitors and became the foundation of our urban climate tool

or might not add texture. While adding texture will contribute richness to the exploration and a more realistic view, if we want to focus on data instead, we can opt for a clean look without the texture.

4   The Making of an Interactive Tool

Seeing the need of an interactive tool to explore engineering and or scientific simulation data and the environment in an urban participatory planning platform, we then develop a tool using the Unity3D game engine. We seek to make the process of designing in the virtual world as iterative and intuitive as it is in the real life to become accepted and utilized by different stakeholders. Design and visualization have been implemented as a loop around and including Unity3D as in Fig. 5, rather than a straight process involving various software elements. In the context of urban design, the urban science simulation result is then integrated as part of decision-making process (Berger et al. 2015). In additional, storytelling principles are integrated by injecting urban science data and walking exploration as the element of playful public participation (Krek 2005).

During the development phase, we cooperated with architects and researcher closely such that we can develop a suited tool for the exploration. It is also aimed that the tool can be used by a general public or lay audience and as such, we noted that there will be adjustment in the features of the tool: i.e., a general public does not need as much detail and information as researchers do. Menu bars and options also could be simplified to avoid confusion.

Fig. 5 Unity3D editor: The creator of the tool is able to import 3D models and change various parameters of the component (*right panel*). The components in the scene are listed on the *left panel*. The *bottom panel* shows the list of files that make up the project (textures, scripts, 3D model, images, audio, etc.)

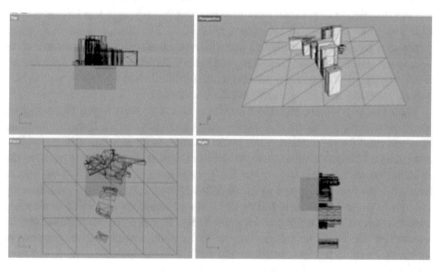

Fig. 6 The 3D model of the city can be imported from various 3D modeling software sources. We simplified the triangles to make the rendering jobs less heavy

To produce such a tool, we first imported the 3D geometry elements and engineering/scientific simulation results into the game engine, as in Fig. 6. The format of the geometry can be of *.obj, *.dae, *.fbx, or other proprietary 3D elements.

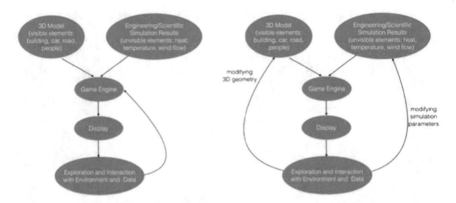

Fig. 7 Simulation–Visualization–Exploration as a flow on the left versus, Simulation–Visualization–Exploration as a design loop. Exploration results could affect the next iteration of the design process, which will then provide new input parameters to the simulation

Currently, the format of the simulation data is any text-based format like *.txt, *.csv. Loading data into RAM memory will provide a better frame rate for interactivity in case (a) sufficient RAM is provided and (b) a balance between visual details and scarcity of file size is kept. The game engine will then render the result and the user is free to explore in the environment using any input controllers that has been setup with the game engine. Several input controllers like mouse, keyboard, game controller (joystick), gesture-based input (Microsoft *Kinect*) have been tested, or even touch-based input and head-mounted input that can track eye/head movement could be used. With the recent advancement of virtual reality gaming, we can also include the treadmill as an input source for positioning.

There can be up to six degrees of freedom in the movement input. Three degrees are employed for positional (up–down, frontward–backward, left–right). Another three are used for orientation: yaw, pitch, and roll, each is rotation in x-, y-, and z-axis, respectively. To have a user perspective when he/she is walking between the buildings or on the road, we use the first-person perspective with five degrees of freedom—the user is allowed to tilt their heads to simulate rotation, and moving frontwards or backwards and left or right. We disabled up and down to simulate realism because the user is not assumed to fly. Elements of playfulness can be injected by allowing vertical movements to a certain extend such as jumping or squatting. Because the simulation mainly deals with outdoor data, we do not have cases such walking up and down the stairs indoor. We still allow another mode where the user freely roams in the 3D environment as if using a helicopter. We adjusted the travel speed to be faster in this exploration mode. The entire procedure is shown in a conceptual flow chart in Fig. 7.

4.1 Study Case: CFDtoUnity3D

In our first study case (Berger and Cristie 2015) we utilize Unity3D to visualize problems of computational fluid dynamics (CFD) to address outdoor thermal comfort (OTC) (Nikolopoulou 2011; Tapias and Schmitt 2014). Simulated are wind flow, temperature, and pressure around a complex of buildings in the University Town area, Singapore. Due to the humid tropical climate urban planning and design is more and more dealing with a loss in the quality of living, perceived as reduced OTC. We use ANSYS as the CFD simulation software. ANSYS divides the simulation space into polygonal elements of the 3D space, the so-called meshing. After numerical calculation of the fluid dynamic behavior, we have data of temperature and pressure within the volume. We import the results based on a rectangular grid with each block defined by its center's position and size.

Wind flow data is independent of blocks and represented in the format of streamlines or tube. For each streamline, we have information about the speed at each specific time step. We then can use this information to create an animation of wind flow by adjusting the wind speed to the frame rate of the interaction tool, from time = 0 s (beginning of animation) until the final time step. Seeing an animation is a bridge for the wider audience to understand more, compared static wind lines. When the wind is hitting the building, we can notice that the speed is going slower. We could also observe how the wind is dispersing, leaving some area to be windier, and some area to be less windy. The tool has been designed to run especially on large and high resolution devices, which is best suitable for workshops with many stakeholders involved. In test cases it runs for small scale on the 5 k resolution iMac, for large scale on our 4 × 4 full-HD screen with 33 megapixel on about 3 × 4 m in the Future City Lab's *ValueLab Asia* (Anwar et al. 2015). According to the large screen, up to 16 sub-screens [each full-HD] could be opened in parallel, however, four sub-screens were regarded as optimal by the test audience. The site of the case study in Figs. 8 and 9 has been originally designed in a way to improve wind flow between buildings [by Perkins and Will architects]. Yet the design process was not public or open to shareholders.

4.2 Study Case: CityHeat

CityHeat is the second study case of a tool where we bring the context of [anthropogenic] heat produced by traffic in a stretch of the Ayer–Rajah Expressway, Singapore (Cristie et al. 2015; Wagner et al. 2015). Here, we would like to invite users to interact with resulting data from heat simulations. To display, we experimented with using a touch screen rather than the video wall. Users can have immediate feedback based on their gesture control on the screen. The interactions include the navigation in the 3D space and time, and also data visualization control (filtering, color settings, output styles).

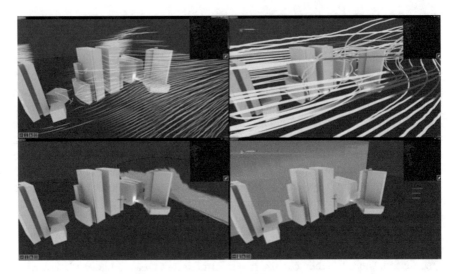

Fig. 8 Multiscreen implementation can also be used to show different representations of the same dataset. The wind flow visualization is shown by animation (*top-left*), static lines (*top-right*), and the wind's wave-front as a triangular surface (*bottom-left*). In additional, by showing the pressure visualization (*bottom-right*), we can see how wind flow and pressure correlates to each other over time

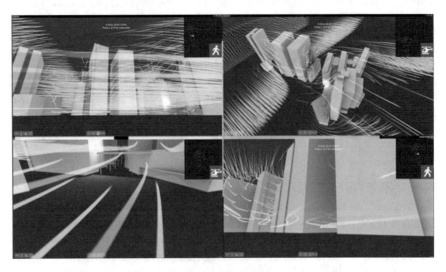

Fig. 9 In a different setup multiscreen implementations allows the user to observe wind flows from different angles to get a better holistic understanding of the scene

Fig. 10 Traffic heat visualization using cubes. More details are shown in the area near the engines of the vehicles. In the picture, *red* color is the cube with the highest temperature, and *blue* color is set as average air temperature

From an engineering point of view, similar to *CFDtoUnity3D*, we also employ animation to help the visualization show how heat is produced from vehicle's components (engine, exhaust, underbody, breaks, tires, and cooling) and its propagation over time. We added a timeline player that has a play and pause button so that the user can stop and move the time-slider to observe the heat propagation at a specific points in time. Additionally, we allow a dynamic color configuration for users to be able to differentiate temperature gradients better. In doing so, one could observe that the air area above the asphalt and below the vehicles will be hotter due to heat from the vehicle through radiation, convection, and friction. As we are using cubes within the 3D cellular automata to visualize the air temperature, we note that too many visible cubes will lead to distracted vision and inability to focus and get a meaning out of the data, as in Fig. 10. On the technical side, it is massively slowing down the rendering process. As such, we had filtering options where user can filter to only see certain cubes with certain range of temperature. We also set a cluster filter where the simulation area is divided into sectors and user can choose the area that they would like to zoom in. Furthermore, we allow users to customize the visibility of vehicles and trees. Disabling vehicles and or trees will allow user to really focus on the zoom in air cubes. On top, an octree algorithm aggregates [and disaggregates] cubes of similar [and different] temperature levels, in order to minimize the number of cubes as much as possible without losing information. This is done already in the scientific simulation level before the data transfer to Unity3D.

Perhaps, one of the most important features in Fig. 11 of this tool is the adjustability of level of details. When enabling automatic level of details, cubes will be less detailed from far, and more detailed as we go nearer to the road level. This is an

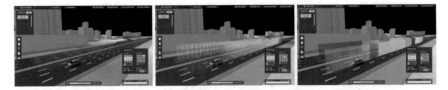

Fig. 11 Gradual level of details visualization. Bigger cubes can be used when viewing from far (*left*), and smaller cubes can be used when we are inspecting closer (*right*)

Fig. 12 A collaborative environment in urban planning that fosters exploration. Each participant/ stakeholder discovers through playing and is enable to engage in further discussion

important timely feedback that will encourage discoverability. Currently, both *CFDtoUnity3D* and *CityHeat* are installed on a local machine due to the huge amount of data imported from the simulation. We could show that the tools are a necessary element in a collaborative environment such that all participants could be engaged in understanding the urban science better, as rendered in Fig. 12. Running seamlessly as normal executable program, the program and relevant data could then just be copied to similar discussion center across the city to get the citizen or other kinds of stakeholders talking. In addition, we could also foresee this tool to be put on a web platform where greater number of audiences could join and participate via remote connections.

5 Conclusions

We have demonstrated how game engines can be used to create exploration tools not only for buildings and for environments, but also introduced how they could act as a bridge to stakeholders who are otherwise not usually connected to urban science, because the data visualized is too complex otherwise. We see urban science as a narrative that emerges during a serious game approach as the data is being explored, in the embedded context of urban area we are exploring. By means of exploration of the 3D space and the data, the tools have introduced knowledge via discovery, be it through personal experience or collaborative exploration.

It is noted that urban planning is an interdisciplinary endeavor, which can be supported by related data about the urban environment. To navigate easily in potentially immense data sets, we have mentioned how data can be segregated according to user needs, and most importantly, we focused on showing enough level of detail to gain understanding without compromising visual or perceptual limitation by displaying too much information or data at once.

In both study cases, we have highlighted the importance of having a first-person perspective and discovery together with the flexibility of user-defined settings. Information panels are also deemed necessary to put visuals into its context.

While looking forward to more similar tools to be developed and easily deployed in the future (in fact, the web is a great platform), we also look to close the loop of design–simulation–visualization by allowing via game engines the user to modify the design and consecutive simulation parameters. In this light, not only everyone can get a scientific story of his or her city, but also everyone then can take part in his or her city planning process.

References

Abt, C.C.: Serious Games. University Press of America (1987)

Ampatzidou, C., Molenda, A.: Building Stories—The Architectural Design Process as Narrative. In: Paper Presented at the Digital Storytelling in Times of Crisis, Athens, Greece. http://www.cristina-ampatzidou.com/building-stories-the-architectural-design-process-as-narrative-conference-paper/ (2014)

Andreoli, R., De Chiara, R., Erra, U., Iannaccone, A., La Greca, F., Scarano, V.: Some Real Experiences in Developing Virtual Environments. In: Paper Presented at the Tenth International Conference on Information Visualization 2006, IV 2006 (2006)

Anwar, A., Klein, B., Berger, M., Müller Arisona, S.: Value Lab Asia: A Space for Physical and Virtual Interdisciplinary Research and Collaboration. In: Paper presented at the 19th International Conference on Information Visualisation (IV 2015) (2015)

Berger, M.: Augmenting science through art. J. Prof. Commun. 3(2) (2014)

Berger, M., Cristie, V.: CFD post-processing in Unity3D. Procedia Comput. Sci. 51, 2913–2922 (2015)

Berger, M., Buš, P., Cristie, V., Kumar, A.: CAD integrated workflow with urban simulation-design loop process. Sustain. City X 194, 11 (2015)

Bodker, K., Kensing, F., Simonsen, J.: Participatory IT Design: Designing for Business and Workplace Realities. MIT Press, Cambridge, MA (2004)

Boeykens, S.: Using 3D Design Software, BIM and Game Engines for Architectural Historical Reconstruction. CAAD Futures, Liege, Belgium (2011)

Brown, T.: Design thinking. Harvard Bus. Rev. 86(6), 84 (2008)

Cecchini, A., Rizzi, P.: Is urban gaming simulation useful? Simul. Gaming 32(4), 507–521 (2001). doi:10.1177/104687810103200407

Chen, I.R., Schnabel, M.A.: Multi-Touch: The Future of Design Interaction. In: Paper Presented at the CAADFutures 2011, Designing Together, Liège (2011)

Clemenson, G.D., Stark, C.E.: Virtual environmental enrichment through video games improves hippocampal-associated memory. J. Neurosci. **35**(49), 16116–16125 (2015)

Cristie, V., Berger, M., Buš, P., Kumar, A., Klein, B.: CityHeat: Visualizing Cellular Automata-Based Traffic Heat in Unity3D. In: Paper Presented at the SIGGRAPH Asia 2015 Visualization in High Performance Computing (2015)

Crowston, K., Prestopnik, N.R.: Motivation and Data Quality in a Citizen Science Game: A Design Science Evaluation. In: Paper Presented at the 46th Hawaii International Conference on System Sciences (HICSS) 2013 (2013)

Dykes, J., MacEachren, A.M., Kraak, M.-J.: Exploring Geovisualization. Elsevier (2005)

ESA: Essential Facts About the Computer and Video Game Industry. Retrieved from http://www .theesa.com/wp-content/uploads/2015/04/ESA-Essential-Facts-2015.pdf (2015)

Fritsch, D., Kada, M.: Visualisation using game engines. Archiwum ISPRS **35**, B5 (2004)

Greenwood, P., Sago, J., Richmond, S., Chau, V.: Using Game Engine Technology to Create Real-Time Interactive Environments to Assist in Planning and Visual Assessment for Infrastructure. In: Paper Presented at the 18th World IMACS/MODSIM Congress, Cairns (2009)

Guhathakurta, S.: Urban modeling as storytelling: using simulation models as a narrative. Environ. Plan. **29**(6), 895–911 (2002)

Gün, O.Y.: Performing Realism: Interactive Real-Time-Rendered Stereoscopic Animations for Architectural Design Process. In: Paper Presented at the eCAADe 2013: Computation and Performance–Proceedings of the 31st International Conference on Education and Research in Computer Aided Architectural Design in Europe, Delft, The Netherlands, 18–20 Sept 2013

Humbert, P., Chevrier, C., Bur, D.: Use of a Real Time 3D Engine for the Visualization of a Town Scale Model Dating from the 19th Century. Retrieved from Prague (2011)

Indraprastha, A., Shinozaki, M.: *Constructing virtual urban environment using game technology.* Paper presented at the 26th eCAADe Conference: Architecture in Computro Antwerpen (2008)

Jenkins, H.: Game design as narrative architecture. Computer **44**(3), 118–130 (2004)

Kim, D.-H., Bejleri, I.: Using 3D GIS Simulation for Urban Design. In: Paper Presented at the ESRI Users conference (2005)

Klein, B., Treyer, L., Zünd, D., Müller Arisona, S.: Value Lab Asia. FCL Magazine **3**, 6–13 (2014)

Krek, A.: Rational Ignorance of the Citizens in Public Participatory Planning. In: Paper Presented at the 10th Symposium on Information-and Communication Technologies (ICT) in Urban Planning and Spatial Development and Impacts of ICT on Physical Space, CORP (2005)

Kühn, W., Jha, M.: Using Visualization for the Design Process of Rural Roads. In: Paper Presented at the 5th International Visualization in Transportation Symposium and Workshop (2006)

Kunze, A., Burkhard, R., Schmitt, G., Tuncer, B.: Visualization and decision support tools in urban planning. Digital Urban Modeling and Simulation, pp. 279–298. Springer (2012)

LeBlanc, M.: Formal Design Tools: Emergent Complexity, Emergent Narrative. In: Paper Presented at the Proceedings of the 2000 Game Developers Conference (2000)

Lévy, P.: Collective Intelligence. Plenum/Harper Collins (1997)

Liu, Z., Jiang, B., Heer, J.: imMens: Real-time Visual Querying of Big Data. In: Paper Presented at the Computer Graphics Forum (2013)

Manyoky, M., Wissen Hayek, U., Heutschi, K., Pieren, R., Grêt-Regamey, A.: Developing a GIS-based visual-acoustic 3D simulation for wind farm assessment. ISPRS Int. J. Geo-Information **3**(1), 29–48 (2014)

Meirelles, I.: Design for Information: An Introduction to the Histories, Theories, And Best Practices Behind Effective Information Visualizations. Rockport publishers (2013)

Müller Arisona, S., Aschwanden, G., Halatsch, J., Wonka, P.: Digital Urban Modeling and Simulation. Springer (2012)

Nikolopoulou, M.: Outdoor thermal comfort. Frontiers in Bioscience **3**, 1552–1568 (2011)

Papadopoulou, M., Raphael, B., Smith, I.F.C., Sekhar, C.: Optimal sensor placement for time-dependent systems: application to wind studies around buildings. J. Comput. Civil Eng. (under review) (2015)

Pettit, C., Widjaja, I., Russo, P., Sinnott, R., Stimson, R., Tomko, M.: Visualisation Support for Exploring Urban Space and Place. In: Paper Presented at the XXII ISPRS Congress, Melbourne (2012)

Poplin, A.: Playful public participation in urban planning: a case study for online serious games. Comput. Environ. Urban Syst. **36**(3), 195–206 (2012)

Prensky, M.: Digital natives, digital immigrants part 1. On the horizon **9**(5), 1–6 (2001)

Salim, F.: Tangible 3D Urban Simulation Table. In: Paper Presented at the Proceedings of the Symposium on Simulation for Architecture and Urban Design (2014)

Schmitt, G., Wenz, F., Kurmann, D., Mark, E.V.D.: Toward virtual reality in architecture: concepts and scenarios from the architectural space laboratory. Presence Teleoperators Virtual Environ. **4**(3), 267–285 (1995)

Seichter, H.: Augmented reality and tangible interfaces in collaborative urban design. Computer-Aided Architectural Design Futures (CAADFutures) 2007, pp. 3–16. Springer (2007)

Signorelli, V.: Game Engines in Urban Planning: Visual and Sound Representations in Public Space. In: Paper Presented at the International Conference Virtual City and Territory (7è: 2011: Lisboa) (2011)

Slowey, K.: New player: How gaming technology adds a 'wow factor' to BIM. Retrieved from http://www.constructiondive.com/news/new-player-how-gaming-technology-adds-a-wow-factor-to-bim/410956/ (2015)

Tan, E.: Play the city. Retrieved from http://playthecity.eu/ (2015)

Tapias, E., Schmitt, G.: Climate-sensitive urban growth: outdoor thermal comfort as an indicator for the design of urban spaces. WIT Trans. Ecol. Environ. **191**, 623–634 (2014)

Treyer, L., Müller Arisona, S., Schmitt, G.: architectural projections: changing the perception of architecture with light. Leonardo electronic almanac **19**(3), 148–163 (2013)

URA (Producer): SingCity. Retrieved from https://www.ura.gov.sg/uol/citygallery?p1=learn&p2=games (2016)

van der Net, M. (Producer): OSCity. Retrieved from http://www.oscity.eu/ (2013)

Vemuri, K., Poplin, A., Monachesi, P.: YouPlaceIt!: A Serious Digital Game for Achieving Consensus in Urban Planning. In: Paper Presented at the AGILE 2014 Castellón, Spain (2014)

Vosinakis, S., Koutsabasis, P., Stavrakis, M., Viorres, N., Darzentas, J.: Virtual environments for collaborative design: requirements and guidelines from a social action perspective. Co-Design **4**(3), 133–150 (2008)

Wagner, M., Viswanathan, V., Pelzer, D., Berger, M., Aydt, H.: Cellular automata-based anthropogenic heat simulation. Procedia Comput. Sci. **51**, 2107–2116 (2015)

Wikipedia: Citizen Design Science. Retrieved from https://en.wikipedia.org/wiki/Citizen_Design_Science (2016)

Winder, I.: CityScope. Retrieved from http://cp.media.mit.edu/city-simulation/ (2014)

Wissen Hayek, U.: Virtuelle Landschaften zur partizipativen Planung: Optimierung von 3D-Landschaftsvisualisierungen zur Informationsvermittlung. vol. 5. vdf Hochschulverlag AG (2009)

Xu, Y., Kim, E., Lee, K., Ki, J., Lee, B.: FDS simulation high rise building model for unity 3D game engine. Int. J. Smart Home **7**(5), 263–274 (2013)

Zanola, S., Fabrikant, S.I., Çöltekin, A.: The Effect of Realism on the Confidence in Spatial Data Quality in Stereoscopic 3D Displays. In: Paper Presented at the Proceedings of the 24th International Cartography Conference (ICC 2009), Santiago, Chile (2009)

Zeng, W., Fu, C.-W., Müller Arisona, S., Erath, A., Qu, H.: Visualizing mobility of public transportation system. Vis. Comput. Graph. IEEE Trans. **20**(12), 1833–1842 (2014)

Part II
Design of Urban and Pervasive Games

Squaring the (Magic) Circle: A Brief Definition and History of Pervasive Games

Eddie Duggan

Abstract Pervasive games defy Johan Huizinga's classic definition of play as being something "outside 'ordinary life'" with their "own proper boundaries of time and space according to fixed rules and an orderly manner". Katie Salen and Eric Zimmerman develop Huizinga's concept of the magic circle and discuss its function as a boundary between the real world and the game world. However, pervasive games seem to form a distinct category of games or types of play that breach both the spatial and the temporal confines of the magic circle. Pervasive games are of particular interest for the way in which they make use of the natural or the built environment as a playspace in a distinct and sometimes alarming overlap with the real world. This chapter offers some definitions and examples of some popular pervasive games, briefly tracing the evolution of treasure hunts, assassination games, live action role-play and alternate reality games, all of which more-or-less confound the notion of the magic circle.

Keywords Alternate reality games · ARGs · Assassination games · Games · LARPs · Letterboxing · Live action role-play · Magic circle · Pervasive games · Playful public performance · Reality games · Role-playing games · RPGs · Smart street sports · Treasure hunt · Urban adventure games

1 Introduction

While Roger Caillois observed that Johan Huizinga omitted to provide a definition of games in *Homo Ludens* (Caillois 2001, p. 4), Huizinga did famously describe play as being "outside ordinary life", as something that "proceeds within its own boundaries of time and space" (Huizinga 1949, p. 13). This chapter will

E. Duggan (✉)
University of Suffolk, Ipswich, UK
e-mail: e.duggan@ucs.ac.uk; e.duggan@uos.ac.uk

© Springer Science+Business Media Singapore 2017
A. Nijholt (ed.), *Playable Cities*, Gaming Media and Social Effects,
DOI 10.1007/978-981-10-1962-3_6

discuss so-called "pervasive games" which blur the boundaries between ordinary life and the ritualistic game space inside the magic circle.

Katie Salen and Eric Zimmerman take Huizinga's notion of the "magic circle" and use the term to describe the physical and temporal boundaries within which play occurs (Salen and Zimmerman 2004, pp. 94–99). Salen and Zimmerman discuss the conceptual boundary around the game as being either "closed", for example, when a game is considered as a set of rules (semiotically, we might call this the *langue* of a game as distinct from any particular player interaction or *parole*[1]) or "open", when a game is subject to a broader approach, for example, a consideration of how meaning is generated in a wider cultural context. They acknowledge that the distinction between "playing" and "not playing" can be "fuzzy and permeable"

> Compare, for example, the informal play of a toy with the more formal play of a game. A child approaching a doll, for example, can slowly and gradually enter into a play relationship with the doll. The child might look at the doll from across the room and shoot it a playful glance. Later, the child might pick it up and hold it, then put it down and leave it for a time. The child might carelessly drag the doll around the room, sometimes talking to it and acknowledging it, at other times forgetting it is there. The boundary between the act of playing with the doll and not playing with the doll is fuzzy and permeable. Within this scenario, we can identify concrete play behaviours, such as making the doll move like a puppet. But there are just as many ambiguous behaviours, which might or [might] not be play, such as idly kneading its head while watching TV. There may be a frame between playing and not playing, but its boundaries are indistinct. (Salen and Zimmerman 2004, p. 94)

However, Salen and Zimmerman's account of the boundaries between playing and not playing does not seem to really fully accommodate the peculiarity of the boundary slippage that can occur in pervasive games.[2]

Jaakko Stenros and Markus Montola define pervasive games as both a "sub-category of games" and "an expansion of what games are" (Montola et al. 2009, p. 31). They discuss some of the key characteristics of pervasive games and group them into two categories, which they call "established genres" and "emerging genres" (Montola et al. 2009, p. 31), along with the caveat that these categories are neither natural nor discovered, but are tentative and provisional generic

---

[1] See Barthes (1964) p. 14 for a distinction between the available speech acts in a language system (*langue*), which Barthes likens to the rules of a game, and any particular speech act (*parole*) which, like a particular instance of a game, is constrained by rules.

[2] Eric Zimmerman revisits the territory of the magic circle in a later online essay written as a rejoinder to those "earnest graduate students" and eminent scholars who would tilt at the windmill of the magic circle as described by Salen and Zimmerman in *Rules of Play* (see Zimmerman 2012). In this later piece, Zimmerman clarifies the position of *Rules of Play*: like a pervasive game, it too simultaneously occupies different spaces—the example he gives is of a work with sociological content that is not itself a sociological text: a better description of *Rules of Play* might refer to it (as, indeed, he does) as an inter- or multi-disciplinary work that doesn't occupy a single position. The present work, albeit by neither an earnest graduate nor an eminent scholar, doesn't seek to denigrate Salen and Zimmermans' concept—or to join "the magic circle jerk" as Zimmerman might put it—but to merely point out the interesting double articulation presented by pervasive games.

groupings created by Stenros and Montola as a way of describing and analysing various aspects of pervasive games.

This chapter, developed from a presentation at the 18th Annual Board Game Studies Colloquium at the Swiss Museum of Games (Duggan 2015), seeks to provide a brief overview of pervasive games and, with reference to several particular examples, to illustrate the blurring of the boundaries of the magic circle in games that use the built environment as game space.

2   Established Genres

Stenros and Montola's "established genres" category consists of existing game types that are likely to be well-known: treasure hunts, assassination games, pervasive LARPs (live action role-play games) and ARGs (alternate reality games). While these four well-known types of pervasive game will be considered briefly in turn, it's worth noting that assassin games (see Sect. 2.2) may be considered an off-the-board form of hunt or chase board games, such as Fox and Geese,[3] with human agents as differentiated pieces on a playing field consisting of the built environment.

2.1   Treasure Hunt

Treasure hunt games probably have their origins in "letterboxing", an English pastime that started in 1854 when Dartmoor guide James Perrott placed a glass jar on a small cairn of stones he had set at a remote spot on the moor. Perrott would invite visitors he had led to the spot, at Cranmere Pool, to place their calling card in the jar for posterity. The visit was no mean feat as the round trip to Cranmere Pool from Perrott's base, a tackle shop at Chagford, involved a difficult sixteen-mile hike across deep tussocks on boggy land.

Perrott's calling-card trek was noted by William Crossing in *Crossing's Guide to Dartmoor* (Crossing 1909). By this time, however, Perrott himself had died (although his sons continued to work as guides), the glass bottle had been replaced by a tin box and the cairn by a stone structure, which came to be called a "letterbox". The practice developed during the first decades of the twentieth century and, in 1937, a more permanent structure, sponsored by the *Western Morning News*, was erected at the site (see Fig. 1). The following year a plaque to commemorate William Crossing was installed at Duck's Pool, the site of another Dartmoor letterbox.

Both sites are marked on the Ordnance Survey Sheet 28 (Ordnance Survey 2010). Making the trip to Cranmere Pool became something of a minor craze as hikers began to leave addressed letters and postcards in the letterbox and the next

---

[3]Fox and Geese: see Murray (1952) pp. 101–112; Bell (1979), v. 1, pp. 76–82; Parlett (1999), pp. 185–204.

Fig. 1 A C20th "letterbox" structure built at the site of Perrott's jar, Cranmere Pool, Dartmoor. http://www.geograph.org.uk/photo/371061 Photograph by Patrick Gueulle (https://creativecomm ons.org/licenses/by-sa/2.0/), licensed for reuse under a Creative Commons Licence CC by-SA 2.0 http://creativecommons.org/licenses/by-sa/2.0/

finder would take the deposited item and post it. A system of rubber stamps and log books evolved during the 1960s, providing finders with a means of recording their encounters with letterboxes as the hobby became established. By the mid-1970s there were about a dozen letterboxes on Dartmoor; now there are thousands.

Letterboxing is not a game per se, but participants have to discover the location of a letterbox in order to deposit a letter or postcard. Sometimes this is by word-of-mouth (WOM) and would-be letterboxers will have to find a WOM group and either attend a physical meeting or join a WOM online forum in order to get access to more-or less-cryptic clues relating to letterbox locations. Other means of participation include working through a published list of letterboxes, such as Janet Palmer's introductory volume, *Let's Go Letterboxing: A Beginner's Guide*, which provides letterbox names and map co-ordinates of their locations (Palmer 1998). Success is its own reward in discovering a letterbox and stamping your logbook or in having a postcard arrive, perhaps posted from some exotic location, some weeks or even months after it was deposited.

For Salen and Zimmerman (2004, p. 95), the boundaries of the magic circle are expressed by reference to the participants: they are recognised by their appearance and/or their actions (the dress of a tennis player, the action of moving a chess piece, the performance, dress and speech of actors in the proscenium space); the place in which the activity takes place (the game board, the playing field, the court of law);

or the time in which the activity occurs (between the formally marked start and finish of a football match, until a win condition such as checkmate or potting a sufficient number of snooker balls such that the opponent cannot score enough points is achieved).

Letterboxing, however, confounds the magic circle somewhat in that participants may be dressed the same as other hikers out on the moor (hiking boots, suitable clothing, backpack, etc.) and both use maps and navigational aids to traverse the same non-delineated physical space. No obvious differences in appearance or actions identify the participants, and no time constraints or win condition will be apparent to observers. However, some secrecy pertains to the location of the letterbox itself—for the player/participant the location is a puzzle to be solved—and, in Huizinga's terms, this hidden location is the "consecrated spot" (Huizinga 1949, p. 10) and WOM groups, where the discovery and content of caches may be discussed, might be seen as something akin to a secret society.

"Geo-caching" is a form of treasure hunt that has evolved from letterboxing. Players use GPS-enabled devices to locate boxes containing trinkets and log books. Participants can select an object from the cache which they might deposit in another cache, or exchange a trinket for one of their own. Trinkets are typically low-value items which nonetheless are of some interest. For example, a small figure such as a toy soldier, a toy car, a key ring, badge, marble or an old casino token. More elaborate objects, known as geo-coins, are also used. These are manufactured as cast or minted objects, and sold to members of the geo-caching community, although some geo-cachers make their own. Many geo-coins are trackable items, stamped or engraved with a unique six-digit code that allows enthusiasts to track the item as it is moved from cache to cache, with finders logging the details in an online database. Like letterboxing, geo-caching is not really a game, although there can be a competitive element. Finds can be accrued, which players log on a global database, along with bragging rights and anecdotes which also provide users with a social focus.

It is not only implicitly accepted that geo-cachers will maintain the secrecy of the location of a cache, but geo-caching websites remind participants to take care to approach a site with caution so as to avoid drawing attention to a location when there is a risk of being observed by non-participants, or outsiders who are often referred to as "muggles".[4] Hence the idea of the magic circle functions here as something that affords some protection and those "in the know" actively work to keep the object of interest inside the magic circle—the geo-cache—hidden from outsiders. In the language of geo-caching, geo-cachers maintain secrecy in order to prevent the cache from being "muggled" (i.e. discovered or interfered with by someone who is not a member of the geo-caching community). The Geocacher's Creed requires players to maintain the "integrity of the game pieces" (Geocacher's Creed, no date).

---

[4]The term "muggle" is derived from the Harry Potter series of children's books by J.K. Rowling (for example, *Harry Potter and the Philosopher's Stone*) where it is used to refer to ordinary mortals who lack magical powers. In a geo-caching context, the term "muggle" is used to refer to a community outsider, someone unfamiliar with geo-caching. See https://www.geocaching.com/abo ut/glossary.aspx.

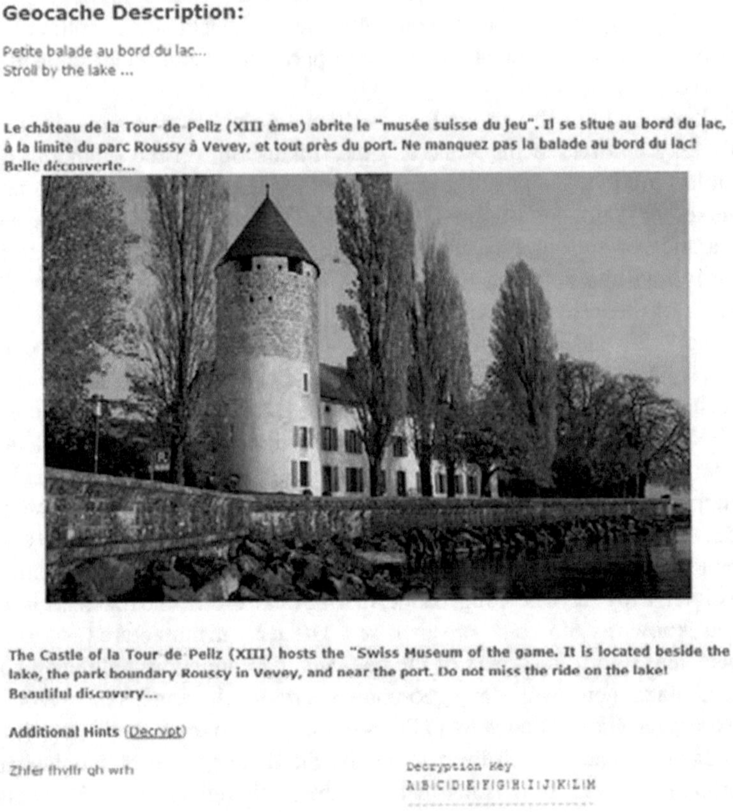

Fig. 2 The Geo-cache website shows a small cache is located at the Swiss Museum of Games. Screenshot: http://www.geocaching.com/geocache/GC2DYPB_vibiscum-ii-chateau-de-la-tour-de-peilz

Geo-caches can be found in urban and rural locations all over the world. For example, there is a cache at the Swiss Museum of Games, located in the small Swiss town of La Tour-de-Peilz (Fig. 2). The cache is possibly unknown to the museum's staff and visitors and to all but a few cognoscenti amongst the passers-by.

Other treasure hunt variations include the scavenger hunt, which can consist of collecting physical objects, or photographs of objects, or photographs of the participants performing certain tasks. The phenomenon of competitive scavenger hunts, in the form of a group of people traversing the landscape and searching the built environment for grail-like objects, was brought to the attention of the cinema-going public in the 1963 epic screwball comedy, *It's a Mad, Mad, Mad, Mad World* starring, amongst a roster of familiar contemporary comedy stars, Spencer Tracey, Ethel Merman, Jerry Lewis, Mickey Rooney, Phil Silvers, Terry Thomas, Peter Falk,

webified storytelling and an urban photo hunt combined!

Game Over!

The game and the results are thoroughly described in these publications:

- **Technical details:** V. Tuulos, J. Scheible and H. Nyholm, Combining Web, Mobile Phones and Public Displays in Large-Scale: Manhattan Story Mashup. *Proceedings of the 5th International Conference on Pervasive Computing*, Toronto, Canada, May 2007. Full-Text PDF (Slides of the presentation)
- **Design process:** J. Scheible, V. Tuulos and T. Ojala, Story Mashup: Design and evaluation of novel interactive storytelling game for mobile and web users. *Proceedings of the 6th international conference on Mobile and ubiquitous multimedia*, Oulu, Finland, December 2007. Full-Text PDF

Game Rules
Instructions for Mobile Players

Fig. 3 The Manhattan Story Mashup site is no longer online, but can be retrieved from the Internet Archive. Screenshot: http://web.archive.org/web/20100504074850/http://www.storymashup.org/

Buster Keaton and The Three Stooges who, by turns, compete and cooperate in a race to find a cache of stolen loot hidden by a robber played by Jimmy Durante.

One well-known example of a photo-collecting scavenger-hunt is "Manhattan Story Mashup", devised in 2006 by Nokia (Fig. 3). Players scored points by collecting photographs to illustrate words selected by other players (photos scored points on submission, and scored again when the word the photograph illustrated was correctly guessed by another player); another way of scoring points was by guessing which of four keywords matched a particular image.

Manhattan Story Mashup consisted, in fact, of several interconnected games for different types of player. Participants who registered as mobile players were given a T-shirt, to make them recognisable to other players, and a mobile phone. The mobile players were encouraged to form small cooperative teams of 2–4 players. Mobile players had to perform two tasks: the first task was to take photographs to illustrate words (nouns) sent to their phones; players had a time limit of sixty to ninety seconds to take a photo once they had selected one of the keywords (nouns) that were constantly being sent to their phones. The second task was to match photographs taken by other players to a list of four keywords. Players scored one point for taking a photo, six points for matching a keyword to a photograph taken by another player and nine points for each of their photos that were guessed correctly by another player.

The nouns sent to the mobile players were extracted from sentences submitted by another set of online players who were participating in the game via the web. The web-based players were given the task of selecting a theme for their story and then creating a story outline using sentences that had been created by other web players; the sentences, which could be tweaked, were then submitted. Nouns were extracted from submitted sentences and sent to the mobile players to be illustrated with a photograph. When a sentence had been illustrated, and corroborated by being guessed correctly by two mobile players, it became available to other web players to use in their story outlines.

The best resulting stories—each story being a "mash-up" or collage of sentences and images created by other players—were then displayed online and also on an electronic hoarding in Times Square. Thus, participants engaged in both cooperative and competitive gameplay; the mobile players used the physical space of Manhattan while the web players participated in virtual space from anywhere in the world in order to create a form of public art or an installation on a screen in Times Square for a live audience of spectators. The results of this playful collaboration temporarily transformed one of the city's electronic advertising hoardings into an amusing and diverting display.

While there is no secrecy involved in Manhattan Story Mashup—it would rather defeat the purpose of an activity explicitly designed to call attention to itself and to the endeavours of those engaged in the activity, as a pervasive game it extends the boundaries of the play activity into the physical world and encourages participants to form cooperative teams in order to compete against other teams, and to engage with the physical world in a different way through the game activities.

2.2 Assassination Games

Assassination games such as "Killer" probably derive from the 1965 Italian film, *La Decima Vittima* or *The Tenth Victim*, which was based on a science-fiction short story by Robert Sheckley, "The Seventh Victim", first published in *Galaxy* magazine in 1953. When the film was released in North America, games of Killer were reportedly played on University campuses in its wake. In an Afterword included in a set of rules for Killer published by Steve Jackson Games in 1981, folklorist John William Johnson recalls the game being played at the University of Texas at Austin in 1966:

> The film was released in December of 1965, and was seen all over the country in a matter of months. It was at this juncture in time that the human hunt game spread from cinema to oral tradition. I remember when the first games were played in 1966 on the campus of The University of Texas at Austin. The game in the oral tradition thus has no single origin as the

multiple copies of the film played in hundreds of theatres all across the country at the same time. (Johnson in Jackson 1992, p. 60)

Johnson treats the game as folklore, circulating in an oral tradition amongst a community or "folk" of students with no knowledge of the film or of Sheckley's original story

> When it began, the players of the game were inspired by the film. Now, fifteen years later, the players do not know anything about its origins. I have interviewed over 200 players [...] and only one game organizer had ever heard of the film. Today's players were only three and four years old when the film was making its rounds in American theatres. Thus the game exists today wholly in the oral tradition. (Johnson in Jackson 1992, p. 60)

In the game, players use soft dart guns (Nerf guns), water pistols or alarm clock "bombs" to assassinate their assigned victim and, as the game takes place without the usual spatial or temporal boundaries it breaks through, or at least blurs, the clear delineation of the magic circle.

Because a killer may lay in wait outside the victim's home, or might telephone the victim's friends or workplace in order to establish his or her whereabouts, there is a considerable overlap of "the real world" and "the game world". "Real people"—non-players—may be asked for game information, or may observe—perhaps even be alarmed by or made suspicious of activity related to the game. Thus the boundaries of the magic circle are distorted by a game that does not take place at a specific time, in a particular location with specific players. For players, excitement is heightened by simultaneously inhabiting two worlds: like letter-boxers and geo-cachers, players are both inside *and* outside the game world at the same time, juxtaposing one world with the other. Overt behaviour is covert game-play. The game is a shared secret.

For Salen and Zimmerman, the conflict that occurs within the magic circle is not real. Their discussion of the magic circle focuses on "an understanding of the arti-ficiality of games, the way in that they create their own time and space separate from ordinary life. The idea that the conflict in games is an *artificial* conflict is part of our very definition of games" (Salen and Zimmerman 2004, p. 94).

However, Huizinga's notion that games are "outside 'ordinary life'", which is central to Salen and Zimmerman's discussion of the magic circle, is challenged by the fact that there are now not only professional video gameplayers (see, for example, Russell 2013) but also professional players of games such as poker and chess, and also professional players of sport. In such a context, "games" are very much part of the "ordinary life" of those who make a living through play, as well as being part of the ordinary lives of so-called "hard-core" gamers who play games habitually and frequently (see Adams 2000). While Salen and Zimmerman refer to the "*artificial* conflict" of a game [their emphasis], the conflict of a game can be very real for, say, the tennis player who may stand to win a great deal of money for success in a tour-nament; the conflict of the game is also very real for the football manager who may lose his job if a team is deemed to have lost too many games, while a chess grand-master is also engaged in a very real struggle, both existential and professional (summed up in the headline "Man Mastered by Machine as Deep Blue Triumphs",

Harston 1997), if not also financial: for example, in the 1997 rematch against IBM's Deep Blue, Gary Kasparov took the runner-up prize of $400,000 while the winning Deep Blue team took $700,000. In a similarly oriented consideration of Salen and Zimmerman's notion of artificiality, Elias, Garfield and Gutschera ask if

> [O]ne should take artificial to mean [...] that the game is being done "for fun" in some sense, not as part of a serious purpose such as making a living [...] But then, is professional baseball not a game while amateur baseball is? (Elias et al. 2012, p. 6).

The disconnect between the magic circle and the real world, between participants in a game of Killer and non-players (people in the real world) is simultaneously what makes the game exciting for players and also what can make it alarming or appear threatening to those who might view the magic circle activities without the benefit of a "frame" to clearly delineate the activities and participants as being something apart from the ordinary life of the non-player or the observer who does not share the lusory attitude of the participants.[5]

Assassins Guilds, as student societies for players of Killer are usually known, exist in a number of British university campuses, including Cambridge, Edinburgh and The University of East Anglia. One emerged at the University of Suffolk (formerly UCS) shortly after the author included references to the game, and screened the film *La Decima Vitimma*, in a Computer Games Design class (Fig. 4).

However, the climate of fear promulgated and maintained in some parts of the world after two aeroplanes were flown into the World Trade Centre in New York in 2001 has made it increasingly difficult to play Killer.[6] The Wikipedia page for the game Assassin includes several references to incidents involving the police, or university disciplinary procedures, or both. The page includes a "security fears" section which reads

> In 2008, the University of Nebraska–Lincoln placed a one-year ban on the Assassin Game. The administrators of the school placed the ban after the police had been called by a person who observed one student bringing a Nerf gun to class. Currently, the University of Texas at Dallas and Loyola University New Orleans can call disciplinary proceedings on a student who engages in a game of "assassin, killer, or variations thereof." (Wikipedia, Assassin (game), no date)

---

[5]"Lusory attitude" is a term coined by Bernard Suits to refer to the state of mind of a player who accepts the rules and conventions of a game. See Suits (2014) "The definition", pp. 36ff. The fact that the lusory attitude of the player might not be shared by the non-player or observer is not explicitly considered in Salen and Zimmerman's discussion of the magic circle, although Sniderman's concept of "the frame", to which they do refer, touches on this point (See Salen and Zimmerman 2004, p. 94).

[6]The events involving the World Trade Centre, the Pentagon and another location (a fourth hijacked aeroplane was brought down, with the loss of all lives on board, before it reached its target) have come to be referred to collectively in popular media discourse as "nine-eleven", a reference to the date (September 11) in the North American date format. Coincidentally, 911 (pronounced "nine-one-one" although the numbers are the same as the date 9/11 or "nine eleven") is also the telephone number of the North American emergency services (cf the British "999").

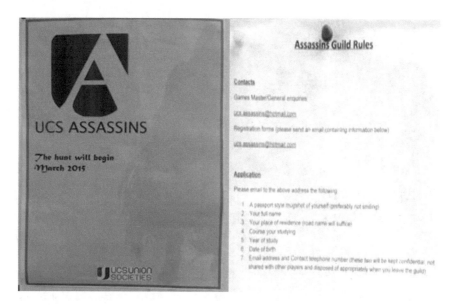

Fig. 4 UCS Assassins Guild. Computer Games Design notice board, University of Suffolk. Photo by the author

Two examples of incidents involving the police, the FBI or other armed response units are also cited

> On March 31, 2009, in Fife, Washington, a Costco, several car dealerships and small businesses were evacuated when a "bomb" was left by someone playing Assassin. Several local police and fire departments responded as well as the Explosives Disposal team from the Port of Seattle, the FBI and the BATF [Bureau of Alcohol, Tobacco Firearms and Explosives]. The bomb was a box "found in a flower bed, contained a magnet and a beeping motion sensor" [*sic*] with the words "bomb, you're dead" written on it. The "bomb" was defused. The man who left the package later turned himself into authorities.
> On May 12, 2009, an incident involving the Assassin game happened behind a North Hampton, New Hampshire, restaurant, where an employee spotted a man in dark clothing with a gun. He called the police, and the student in question [*sic*] did not resist but simply walked to his car and explained the game to the police [*sic*]. The man turned out to be a high school senior from Exeter, New Hampshire, waiting for another high school student to come out of her job at the restaurant with a water pistol in hand [*sic*]. (Wikipedia, Assassin (game))

Jane McGonigal and Ian Bogost devised a Killer-variant, called "Cruel to be Kind" (Fig. 5) which works like a verbal form of rock-paper-scissors, with players offering a greeting ("Welcome to San Francisco"), paying a compliment ("You look gorgeous") or blowing a kiss (McGonigal 2011, pp. 189–197). This variant, subtitled "a game of benevolent assassination", was devised to increase the number of "positive social interactions" in a particular space by utilising game mechanics that consist of greetings and compliments; it also circumvents the restrictions related to the overt form of the assassin game and, at least in the early stages of the game when teams

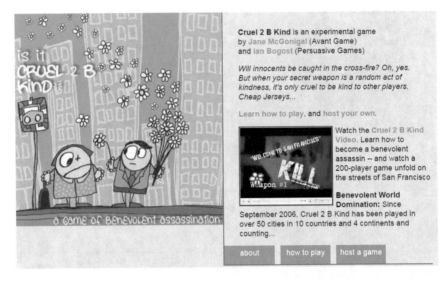

Fig. 5 "Cruel to be kind: a game of benevolent assassination" by Jane McGonigal and Ian Bogost. Screenshot: http://www.cruelgame.com/

are small, avoids causing any alarm to by-standers who might encounter the game in those moments when the game world and the real world collide.[7]

In McGonigal and Bogost's benevolent version of the game, players approach strangers and use one of the weapon-phrases, such as "Welcome to San Francisco", which function in a double articulation to be both an innocuous greeting or innocent but benevolent comment to non-players while being simultaneously recognised as part of the game to participants who are "in on" the secret. Thus the sanctity of the magic circle is preserved, along with non-players' personal sense of well-being, while game participants are still able to enjoy the shared secret of the game. Non-players' sense of well-being may even be enhanced by the experience of a positive social interaction such as a friendly greeting or a benevolent comment from a stranger. It is worth noting that while the game starts as an assassin-game variant, it can develop into a form of playful public performance (see Sect. 3.2) as teams grow in size, resulting in a flash mob type spectacle of a large group of people moving *en masse* through a city space, calling out greetings and compliments to passers-by.

---

[7]McGonigal notes "being accidentally 'attacked' by a player is somewhat startling–but also potentially enjoyable. In a best case scenario, the 'victims' of play feel genuinely welcome or complimented or appreciated. At the start of the game when players are timid and groups are small, this tend to be the case. Later, as players get bolder and teams get larger, strangers are more likely to be clued into the unusual nature of the activity and provoked to wonder why everyone is making such showy efforts of gratitude and kindness." (McGonigal 2011, p. 194).

2.3 Pervasive LARPs

Live action role-play games (LARPs) are a form of improvised performance which, according to Stenros and Montola (Montola et al. 2009, p. 36), evolved from tabletop role-playing games (RPGs) in the early 1980s. Players started to personify or embody game characters, which were typically represented in tabletop games with 25 or 28 mm miniature models. In LARPs, however, players don costumes and, in so doing, they *become* game characters and thus take the game off the board and out into the real world. LARPs also have some overlap with improvised theatre and with Augusto Boal's invisible theatre.[8] Nevelsteen notes a relation between LARPs and MUDs (a Multi-User Domain or Multi-User Dungeon) as observed by Jennica Falk who described MUD as "a virtual counterpart to the LARP" (Nevelsteen 2015, p. 42). Stenros and Montola suggest that LARP games have had less media coverage than some of the other categories of pervasive game and this may explain why LARPs might be considered a lesser known form of pervasive game (Montola et al. 2009, p. 36).

An amusing example of the overlap between the game space and the real world can be seen in *Ikea Heights* (Fig. 6), an episodic performance piece or mini-soap-opera, performed and filmed in a branch of the Ikea furniture chain in Burbank, California, without Ikea's knowledge or permission, at least in early episodes. Ikea staff, when alerted to the goings-on, can be seen sometimes interrupting filming and ejecting the cast and crew from the premises. Customers can be seen and heard in the background, and sometimes the foreground. The series, which consists of seven episodes each with a running time of five minutes, can be found online https://vimeo.com/channels/ikeaheights.

The humour in *Ikea Heights* is generated through the overlap of worlds: the magic circle of the performance space, superposed onto the shop floor of a department store, creates a disjunction when non-participants (either online viewers watching the edited video, or shoppers witnessing the live performance) realise that the performance is being enacted in a requisitioned set that is a re-purposed furniture display simultaneously fulfilling its original function. If the overlap or disjunction were not in evidence—if the performance were staged after hours, and there were no shoppers or shop staff to "intrude"—the piece would still be amusing but not *as* amusing as the heightened effect created by the disjunction of the *Ikea Heights* cast and Ikea shoppers and staff colliding in a juxtaposition of the sacred space of the magic circle and the quotidian space of a furniture shop. Thus the disruption of the magic circle can be seen as an essential aspect of this hybrid form of performance/role play.

In gameplay terms, LARPs are primarily structured around what Caillois would call mimicry (Caillois 2001, p. 23) although *Ikea Heights* is also an example of guerrilla film-making. Stenros and Montola suggest that players of assassination

---

[8]"Invisible Theatre" takes place in a public space among an "audience" of by-standers, who might be drawn into a dialogue with performers, unaware that a performance is being staged. See Boal (2002).

Fig. 6 *Ikea Heights* Episode 1. Screenshot: https://vimeo.com/channels/ikeaheights

games such as Killer could also be said to be LARPing on the strength of their role-playing performances (Montola et al. 2009, p. 35).

2.4 Alternate Reality Games

Alternate reality games (ARGs) are collaborative rather than competitive; groups of players work together to solve puzzles or unravel mysteries. Several well-known examples of ARGs have been used as marketing devices to promote films, television programmes, mobile phones and computer games.

One notable example of an ARG is an unnamed game that came to be known as "The Beast", which was devised by a small team working in the Microsoft Game Group as part of the marketing campaign for the 2001 film *Artificial Intelligence: AI*. The game was designed to provide clues across a variety of media forms, including specially created websites, film posters, film trailers, messages on telephone answering machines and emails, encouraging players to collaborate in solving a puzzle that presented itself as something that was not a game.

The game used the concept of the "rabbit hole" as an entry to the "wonderland" of the game. One example of such an entry point is what Jane McGonigal has described as "a provocative credit": "Jeanine Salla, Sentient Machine Therapist" (McGonigal 2003, p. 2). Anyone sharp eyed enough to spot the credit and inquisitive enough to follow it up, would find planted information, including biographical material and contact details, and links to a website, apparently created by a friend of

Jeanine Salla, which related the tale of a mysterious death of another character, Evan Chan, all set in the future in 2142. Thousands of users collaborated in solving puzzles and interacting with the game.

3 Emerging Genres

Stenros and Montola identify what they call "emerging genres" in the form of reality games, playful public performances, urban adventure games and "smart street sports" (Montola et al. 2009, pp. 40–45). The following section will discuss several examples of pervasive games that conform to Stenros and Montola's "emerging genres" category in terms of the concept of the magic circle.

3.1 Reality Games

Reality games tend toward *paideiatic* rather than *ludic* activities—that is, they constitute a form of play rather than a type of game—something to intrigue or engage bystanders or non-players. Stenros and Montola define reality games as a form of "aesthetic vandalism ... making the urban setting more playful and anarchic" (Montola et al. 2009, p. 44). An example of a reality game overlapping with the real world can be found in the set of Mario Question Blocks "installed" by a group of girls in Ravenna, a small town in Ohio, as an April Fool prank in 2006 after reading a web page with instructions by Toronto-based street artist, Posterchild (see Posterchild 2006).[9] However, the Question Blocks in Ravenna (Fig. 7) were mistaken for a potential terrorist threat by authorities unfamiliar with Super Mario games. This is another example of the concerns that McGonigal and Bogost managed to circumvent with their "Cruel to be Kind" version of "Killer" (see Sect. 2.2). Canadian comic artist Ryan North noted in a comment on Posterchild's webpage that the boxes were intended to "bring a smile to people's faces, to get them to connect with the neighbours, to bring colour into an otherwise grey urban landscape" (Posterchild 2006).

The collision of worlds in Ravenna is a good example of how frame slippage can cause real discomfort: while there may be some amusement in the idea that a certain sector of society—the "straights" or "squares" who are unfamiliar with the iconography of the gamer world and are therefore unable to recognise a Question Block for what it is—who "don't get" the game or the event that is being staged around them, there is a hint of a sneer implied in the tone of some of the newspaper reports and online discussions that seem to address an audience familiar with gaming culture and the iconography of the Nintendo game world (for example, Block 2006, online;

---

[9]"Question Cubes" are game items that appear in the *Super Mario Brothers* series of console games published by Nintendo. Question Cubes typically contain items such as mushrooms, flowers or coins, which can help a player make progress through a game level.

bOiNGbOiNG / CORY DOCTOROW / 1:50 AM THU JUN 9, 2005

HOWTO make life-size Mario power-up blocks

This page documents the project of an artist in Windsor, Ontario, to hang classic Super Mario Brothers "Power-Up Blocks" from trees, power lines, and other high places in his town. He also set out many mushrooms, gold coins and fire-flowers. Link (*via Waxy*)

Fig. 7 How to make life-size Mario power-up blocks Screenshot: http://boingboing.net/2005/06/09/howto-make-lifesize.html

Doctorrow 2006, online). Any sense of fear, panic or sense of danger experienced by those outsiders who are not "in the know" nonetheless warrants empathy because the concern or discomfort experienced by those outside the magic circle is a very real discomfort indeed. Again, we can see how McGonigal and Bogost's strategy to lessen the discomfort caused by frame slippage is a solution that both preserves the frisson for the cognoscenti while simultaneously safeguarding non-participants from experiencing distress when the magic circle and the real world come into collision.

3.2 Playful Public Performance

This type of pervasive game is akin to something from the television series *It's A Knockout (Jeux Sans Frontiere)*:[10] a public spectacle with a performance element. Stenros and Montola cite The Big Urban Game by way of an example. This game was initiated by The University of Minnesota over five days in September 2003. The purpose of the game, designed by Frank Lantz, Katie Salen and Nick Fortugno, was to raise awareness of the urban environment and to encourage participation in discussion of local urban design in the neighbouring cities of St Paul and Minneapolis. The game took the form of a race in which three teams, each with a different starting point, moved a giant game piece along a designated route, selected by online participants, to a bridge linking the two cities. While the magic circle here is the space in which the participants engage with the game's objectives and game pieces, it also functions as a marker to separate participants from non-participants although

[10]See It's a Knockout http://www.its-a-knockout.tv/ for a history of the British and European television series.

it lacks the added frisson that comes from the collision of worlds evident in assassin games or in the guerrilla film-making performance of *Ikea Heights*.

3.3   Urban Adventure Games

Urban Adventure games have been described as the descendants of interactive fiction, combining narrative and puzzle elements with an urban landscape. Stenros and Montola describe urban adventure games as akin to "hypertexts manoeuvred in physical space" (Montola et al. 2009, p. 42). One of the examples they discuss (Case Study K, pp. 215–218) is REXexplorer, designed as an interactive tourist guide for the German city of Regensburg for an audience of 16–30-year-olds. Visitors borrow hand-held mobile devices from the tourist office and use them to complete tasks, solve puzzles and upload photographs and video clips as they explore the city (Ballagas et al. 2006). Data is saved to a blog which participants can access to review their game progress and to remember their visit.

While this type of application can enrich the visitors' experience in a touristic or heritage context, there is no covert activity in this particular use-case scenario. However, an interesting example of an urban adventure game can be found in "The Jejune Institute", an interactive narrative experience that ran in San Francisco from 2008 to 2011. Something of an art installation-*cum*-puzzle-*cum*-treasure hunt, The Jejune Institute provided participants with an enigmatic immersive experience based around the story of a missing girl and a mysterious cult.

The designer/producer of The Jejune Institute is Jeff Hull, a San Francisco Bay Area artist who is described in his Linkedin entry as a "Situational Designer". In a TEDx SoMa video, Hull outlines his rationale and philosophy on creating playfulness in urban spaces (Hull 2010, online). Hull recalls that in the Fairyland Theme Park in Oakland (where Hull worked as a child performer) one could literally go down a rabbit hole (or tunnel) to discover sights and sounds, dioramas and stories to encounter and interact with the environment in ways that differed from the ordinary everyday experience. Hull seeks to bring something of that sense of discovery and play to the everyday world that exists between home and work, a realm that sociologist Ray Oldenburg refers to as the "third place" (Oldenburg 1989, pp. 20–42), in order to distinguish it from the first place (home) and the second place (work).[11]

Hull's work in pervasive games developed via a loose collective of artists on the underground scene in Oakland from which evolved the Oaklandish community development project.[12] For Hull, Oldenburg's third place is essential for developing a sense of community but privatisation and fears about "security" have eroded the

---

[11]Ferreira et al. (2017) discuss Oldenburg's "third place" in this volume; see especially Sect. 2.

[12]See http://oaklandish.com/about for a brief resume of the development of the Oaklandish community project. Greenberg (2015) offers a more detailed account of the emergence of the Oaklandish collective in 2000 and Hull's relationship with it until his departure in 2006 following the closure of the Oaklandish Gallery.

ways in which people can utilise these third spaces. This erosion is evident too in the need for McGonigal and Bogost's assassin variant (see Sect. 2.2).

Apart from playgrounds, which are for children, and sports centres, which are for organised sport, there are few places where adult play can take place. Hull suggests it is difficult for people to engage in playful behaviour after adolescence. While artists and designers are increasingly utilising public space—essentially giving it back to the public for adult playfulness through pervasive games, public art and events such as flash mobs—these are but temporary interventions.

Hull's stated aim is to find way to embed playfulness into the landscape and into the fabric of the environment in more permanent way, to instal rabbit holes in the everyday world. It was this which led to the formation of Nonchalance, Hull's "situational design agency"[13] and to the interactive narrative adventure of The Jejune Institute. Fliers posted around San Francisco in 2008 provided rabbit holes for those curious enough to investigate the material further by telephoning the number given on the flier.

A recorded message gave callers the address of an office building in the city's financial district. Once there, participants would be given a key by a receptionist allowing them to access an office with a video relating a fictional narrative of a mysterious cult which, in turn, led the participant outside to discover further clues embedded in the environment (clues in caches, on statues and plaques, via keys to locked boxes, recorded telephone messages, special radio broadcasts, passwords to secret websites, etc.), all of which provide or encourage different ways of interacting with the built environment. In all, Hull created five different chapters or adventures, each set in a different district of San Francisco. According to *The Wall Street Journal* (Woo 2011) more than 7000 participants engaged with the first chapter of The Jejune Institute.

While the Jejune Institute did indeed enable participants to undertake a form of urban exploration that redefined the urban space as a playful space, *The Institute*, a 2013 documentary about the undertaking, or perhaps a "mockumentary" that presents itself as a documentary utilising the "this is not a game" trope, prompts the viewer to question the veracity of the material itself: something akin to the "is it real or is it a spoof?" effect exemplified in the 2010 film, *Exit Through The Gift Shop*.

In *Exit Through the Gift Shop*, the viewer is left wondering if the character Thierry Guetta is real, or if he, and the story within the film, is actually an elaborate spoof or an art-prank devised by the graffiti artist Banksy to confound the narrative and subvert the film's ostensible documentary form.[14] In Spencer McCall's *The Institute*, the result is less than satisfying as the film seems to be a spoof rather than a reliable account of the Jejune Project itself. This is caused primarily by the conceit adopted by McCall to present the documentary as being the result of his trying to "unravel the mystery" of the Jejune Institute (Gryniewicz 2014, online). This framing device, what we might call the "playfulness" or game element of what should be,

---

[13]See the Nonchalance mission statement http://nonchalance.com.

[14]See Walker (2010) for an interpretation of *Exit Through the Gift Shop* as an elaborate spoof. Cf Miller's reading in *Sight & Sound* (see Miller 2010).

according to documentary conventions, a factual film, serves to undermine not only McCall's film, but also prompts the viewer to doubt the veracity of Jeff Hull's entire project, and to consider that The Jejune Institute may be no more than a fictional element in a spoof documentary.

The effect created by the film, or the present author's reading of it, may be unfortunate as such a response detracts from the nobler intentions evident in Hull's philosophy and in the very real social change that has been wrought and, indeed, continues through the Oaklandish project. It is only through a degree of tenacity, *pace* McCall, in really "unravelling" the story behind the film and the Jejune Institute that the authenticity of Hull, Nonchalance, the Jejune Institute and Oaklandish, was established. It may simply be that in seeking to extend the reach of the Jejune Institute from a location-based experience to film, McCall had not made a satisfactory or sufficiently permeable bridge between the real world and the magic circle.[15]

Urban Adventure Games differ from Playful Public Performance because they lack the performative element (although the game masters of The Jejune Institute and the game characters who appear in video clips and in person take on a performative role even if the player-participants do not); these games have a more relaxed pace than the frenetic Smart Street Sports to be considered below. However, it is worth reiterating that Stenros and Montola's categories are not fixed and certain, but fluid and tentative. That said, a Google search for "urban adventure game" suggests that the term has been appropriated for community-based sports activities with a charitable aspect (for example, community events involving teams competing in running, cycling, swimming, etc. as a means of fund raising).

3.4   Smart Street Sports

The "smart" in "Smart Street Sports" is provided by technology. The form tends to consist of re-creations of computer games in physical space, or technologically enhanced "tag"-style games. Stenros et al. observe the arcade game *PacMan* —essentially a chase game in which the player character, known as PacMan, either runs away from or pursues four ghosts—has inspired several Smart Street Sport re-creations (Montola et al. 2009). PacManhattan is a well-known example of such an implementation. Set in the street grid around Washington Square Park in the Greenwich Village area of Manhattan, the game was created as a postgraduate project by a group of students at New York University in 2004.

An earlier real-world implementation of *Pac-Man* was created by Adrian Cheok in the Mixed Reality Lab at the University of Singapore in 2002 (Cheok et al. 2004).[16] Entitled Human PacMan, the aim of Cheok's game was to allow players freedom to

---

[15]According to IMDB, a sequel of sorts, *The Esquire*, is scheduled for release in 2016. *The Esquire* tells the story of Octavio Coleman, the leader of the cult in The Jejune Institute.

[16]I am grateful to Anton Nijholt for directing my attention to an even earlier real-world implementation of a computer game in the form of ARQuake: see Thomas et al. (2000).

move around in both the real world and the game world, interacting with both simultaneously, thereby overcoming the limitations of arcade games and console games such as *Dance Dance Revolution* which, while combining physical activity with recreational computing, constrain players to a specific area such as a dance mat (Cheok et al. 2004, pp. 71–72).

In Human PacMan, players carry a backpack of hardware containing a laptop, battery, touch sensor, wireless LAN card and other components, and wear a head-mounted display (HMD) in order to view the game world, a combination of the real world and an Augmented Reality layer. Bluetooth technology detects physical contact with the backpack and player interaction with in-game items ("pellets") and with other players. Head-motion technology monitors the position of the player's head while GPS tracks the player's position in space (Cheok et al. 2004, p. 73).

Players are simultaneously in the real world and a virtual world. The aim for the PacMan players is to collect all the "pellets" placed in the real world ("pellets" are real objects containing Bluetooth devices); when a PacMan player is close to a pellet, an alert is triggered in the HMD, indicating a pellet is nearby. The player then locates the object in the real world and picks it up; motion detection causes the pellet to be added to player's inventory. Ghost players, meanwhile, seek to capture the PacMan players by touching them on the shoulder triggering the touch sensor. This game of electronic "tag" is made trickier as the Ghost players cannot see the PacMan players; they can, however, see the pellets disappear in the HMD as they are collected by the PacMan players (Cheok et al. 2004, p. 75).

Each in-game player has a Helper in the form of a human who can "see" all the players and objects. Helpers can interact with each other in order to help their players (PacMan or Ghost players) to achieve their objectives, such as giving directions to help a Ghost capture a PacMan player or directing a PacMan player toward any remaining pellets.

Cheok et al. remark that user feedback suggested players found Human PacMan more entertaining than the *PacMan* computer game, but the heavy equipment needed in the backpack for the wearable computers, and the cumbersome HMDs which provided players with an Augmented Reality layer, were found to be drawbacks, as was the accuracy of the location sensing devices which tended to cause players to "drift". Players also found the Helpers confusing (Cheok et al. 2004, p. 77). It should be noted too that funding for the project was provided by the Defence Science and Technology Agency of Singapore so the apparent benevolence of a technologically enhanced game takes on a more sinister dimension when it serves as a test-bed for militaristic activity (Mixed Reality Lab, no date, FAQ#16), particularly in a repressive regime with a poor record relating to human rights (see, for example, Human Rights Watch 2015, online).

In PacManhattan (Fig. 8), each of the five players uses a mobile phone to keep in touch with a team of five controllers who update their player's position on an online map. The PacMan player touches lamp posts to simulate eating pellets while four ghost players try to catch him. When the PacMan player touches a lamp post designated as a "power pill", which allows him to turn the tables and to pursue the ghosts, the PacMan controller tells the PacMan player and the four ghost controllers about

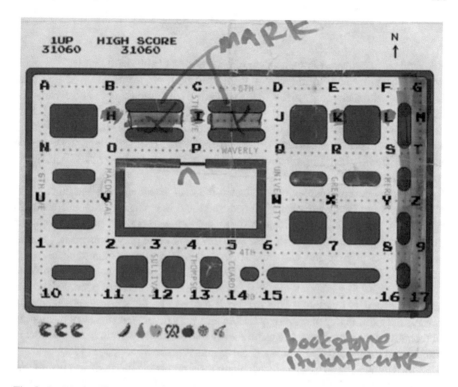

Fig. 8 PacManhattan: play test game map. Image retrieved via the Internet Archive. Screenshot: http://web.archive.org/web/20141011113813/http://www.pacmanhattan.com/about.php

the change in the game state and the four ghost controllers, in turn, relay the information to the four ghost players.

Mobile phones were used for voice calls because the developers found GPS did not work well in the deep valleys of New York's urban environment; the developers were also unable to find an affordable means of handling the GPS data. The original game attracted a lot of attention and was widely covered in the media and online. The game was played again in Brooklyn in July 2014, using the same technology as 2004, as part of the annual Come Out and Play festival. The festival, mostly a New York-based event (although it has been staged in Amsterdam and San Francisco), promotes various types of games and play

> Street games, pervasive games, new urban games, big games, locative games, location-aware games, location-based games, GPS games, flash mob games, augmented reality games, scavenger hunts, art-sports, and even LARPs are all among the diverse kinds of games that can be played at the annual Come Out and Play Festival. (Come Out and Play 2009)

4 Conclusion

It is evident that interest in games and play is in rude health, with innovative developments bringing new forms of play to increasingly larger groups of participants. Innovation with technology and design is enabling both convergence and divergence as new game types are developed and old games are re-created: for example, letter-boxing has evolved from a little-known pastime for visitors to Dartmoor to a global technology-enabled treasure hunt in the form of geo-caching (Sect. 2.1); arcade games are being played in the street with human players acting out the part of the game avatars in real time (Sect. 3.4).

The prevalence of smart-phones, tablet computers, smart watches and other wearable technology (fitness trackers, 3D headsets) and inexpensive devices for developers such as the Raspberry Pi and Arduino continues to make technology-enabled pervasive gaming more accessible for both developers and players as portable hardware becomes lighter and wearable smart technology more commonplace, overcoming the limitations noted by Cheok, for example. However, as Nevelsteen notes (Nevelsteen 2015) games engines that support the development of pervasive games are not yet widely available. This may well prove to be the step-change needed to resolve the gallimaufry observed by Leaver and Willson in their survey of the variety of technologies, game types and terminology that currently pertain in casual, social, mobile and location-based gaming (Leaver and Willson 2016, pp. 1–11).

Both Human PacMan and PacManhattan demonstrate the convergence of the sedentary and the physical, although it can be seen that funding sources for research can introduce a sinister aspect to the usual benevolence associated with game play. For example, Jeff Hull's well-intentioned philosophy, evident from his involvement in the founding of the Oaklandish community development project and his TEDx-SoMa talk (Sect. 3.3), Posterchild's sense of fun, and Jane McGonigal and Ian Bogost's reconfiguration of Killer to circumvent both the administrative repression and the risk of adverse reaction and real distress that can bedevil assassination games (Sect. 2.2) all contrast markedly with the societal impact that might be derived from the outcomes of research that enables what Louis Althusser termed "Repressive State Apparatuses" (see Althusser 1971, pp. 141–148) in order to control or constrain free forms of expression in, say, a demonstration or in public assembly.

Moreover, the very real conflict that can be experienced by those for whom play is an ordinary, everyday undertaking (such as pro-gamers and hard-core gamers) shows the limitations of the formulations of both Huizinga and Salen and Zimmerman, which posit play as a form of artificial conflict that is "outside" ordinary life.

Nonetheless, the wide range of interactive and immersive experiences that overlap in some way with the broadest senses of "game" and "play" demonstrate divergence as more benevolent playful experiences take on new forms (technology-enabled treasure hunts, collaborative ARGs, consciousness-raising through playful public performances, playful installations). The term "jerk", as used in physics to mean a change in acceleration being applied over time (Gibbs 1996), may well be used, *pace*

Zimmerman (2012) (see note 2, Sect. 1) to describe the effect that the various types of pervasive game reviewed here are having on game culture in the playful cities of the late twentieth and early twenty-first centuries.

References

Adams, E.: Casual versus core. Gamasutra, August 1. http://www.gamasutra.com/view/feature/13 1529/casual_versus_core.php (2000). Accessed 10 Apr 2016

Adams, E., Barry, I.: From casual to core: a statistical mechanism for studying gamer dedication. Gamasutra, June 5. http://www.gamasutra.com/view/feature/131397/from_casual_to_core_ a_statistical_.php (2002). Accessed 10 Apr 2016

Althusser, L.: Ideology and ideological state apparatuses (notes toward an investigation). In: "Lenin and Philosophy" and Other Essays, pp. 127–186. Monthly Review Press, New York (1971)

Anon: Girls attempt real-life version of video game. Akron Beacon Journal Ohio, April 1. [Internet Archive] https://web.archive.org/web/20060615102722/ http://www.ohio.com/ mld/ohio/news/14239923.htm (2006). Accessed 20 Mar 2015

Ballagas, R., Walz, S.P., Borchers, J.: REXplorer: A pervasive spell-casting game for tourists as social software. CHI 2006 Workshop on Mobile Social Software (MoSoSo), April 2006. https:// hci.rwth-aachen.de/materials/publications/ballagas2006a.pdf (2006). Accessed 20 Apr 2016

Barthes, R.: Elements of Semiology. Hill & Wang, New York (1964)

Bell, R.C.: Board and Tables Games from Many Civilizations. Dover Publications, New York (1979)

Big Urban Game: University of Minnesota. [Internet Archive] http://web.archive.org/web/200 31208184053/ http://design.umn.edu/bug/index.jsp (2003). Accessed 20 Mar 2015

Bitka, J.: Hell local: Oaklandish spreads the love here and elsewhere. Oaklandish Magazine. http:// www.oaklandmagazine.com/Oakland-Magazine/December-2011/Hella-Local/ (2011). Accessed 1 Apr 2016

Block, R.: Mario question cube girls let off. Engadget, April 8. http://www.engadget.com/2006/0 4/08/mario-question-cube-girls-let-off/ (2006). Accessed 1 Apr 2016

Boal, A.: Invisible Theatre. In: Schneider, R., Cody, G. (eds.) Re-direction: A Theoretical and Practical Guide, pp. 112–124. Routledge, London (2002)

Caillois, R.: Man, Play and Games. University of Illinois Press, Chicago (2001)

Cheok, A.D., et al.: Human Pacman: a mobile, wide-area entertainment system based on physical, social and ubiquitous computing. Pers. Ubiquit. Comput. **8**, 71–81 (2004). doi: 10.10 07/ s00779-004-0267-x

Come Out and Play: Festival Fact Sheet. [Press Release] http://www.comeoutandplay.org/press _factsheet.php (2009). Accessed 20 Mar 2015

Crossing, W.: Guide to Dartmoor: a topographical description of the forest and commons by William Crossing. With maps and sketches. Western Morning News Company Ltd, Plymouth (1909)

Doctorrow, C.: Bomb squad called out to "defuse" life-size Super Mario power-ups. Boing Boing, April 4. http://boingboing.net/2006/04/04/bomb-squad-called-ou.html (2006). Accessed 25 May 2015

Duggan, E.: Off the board: a brief definition and history of pervasive games. In: Paper Presented at the 18th Annual Board Game Studies Colloquium, Swiss Museum of Games, La Tour-des-Peilz, 15–18 Apr 2015

Elias, G.S., Garfield, R., Gutschera, K.R.: Characteristics of Games. MIT Press, London (2012)

Ferreira, V., Anacleto, J., Bueno, A.: Designing ICT for Thirdplaceness. In: Nijholt, A. (ed) Playable Cities: The City as a Digital Playground, pp. 00–00. Springer, Singapore (2017)

Geocachers' Creed: Groundspeak. http://support.groundspeak.com/index.php?pg=kb.page&id=46 (no date). Accessed 10 Mar 2016

Gibbs, P.: What do you call the rate of change of acceleration? The Original Usenet Physics FAQ. http://math.ucr.edu/home/baez/physics/index.html (1996). Accessed 20 Apr 2015

Greenberg, A.: Mystery, commerce, local pride in that one familiar tree. Oakland North. https://oaklandnorth.net/2015/02/24/mystery-commerce-local-pride-in-that-one-familiar-tree/ (2015). Accessed 25 May 2016

Gryniewicz, J.: Neither here nor there: the institute, the game and the thread to elsewhere. Pop Matters, October 13. http://www.popmatters.com/column/186575-neither-here-nor-there/ (2014). Accessed 25 May 2016

Hanke, J.: Reality as virtual playground. Making Games. http://www.makinggames.biz/features/reality-as-a-virtual-playground-7286.html (2015). Accessed 20 Apr 2016

Harston, W.: Man mastered by machine as deep blue triumphs. The Independent, May 11. http://www.independent.co.uk/news/man-mastered-by-machine-as-deep-blue-triumphs-1261087.html (1997). Accessed 10 Apr 2016

Hodson, H.: Why Google's ingress game is a data goldmine. New Scientist. November 28. https://www.newscientist.com/article/mg21628936.200-why-googles-ingress-game-is-a-data-gold-mine/ (2012). Accessed 20 Apr 2016

Hon, A.: The Guide X: A Tale of the A. I. Trail. Cloudmakers. http://www.cloudmakers.org/guide/ (2001). Accessed 10 Mar 2015

Huizing, J.: Homo Ludens: A study of the play element in culture. Routledge & Kegan Paul, London (1949)

Hull, J.: Variability and play in the civic realm. TEDx SoMa. https://www.youtube.com/watch?v=UsyybmZFmXg (2010). Accessed 25 May 2016

Human Rights Watch: World Report 2015: Singapore. https://www.hrw.org/world-report/2015/country-chapters/singapore (2015). Accessed 1 Apr 2016

It's A Knockout: http://www.its-a-knockout.tv/ (no date). Accessed 8 Mar 2016

Jackson, Killer: The Game of Assassination. Steve Jackson Games, Austin (1992)

Leaver, T., Willson, M.: Social networks, casual games and mobile devices: the shifting contexts of gamers and gaming. In: Leaver, T., Willson, M. (eds.) Social, Casual and Mobile Games: The Changing Gaming Landscape, pp. 1–11. Bloomsbury, London (2016)

McGonigal, J.: 'This is not a game': Immersive aesthetics and collective play. In: Digital Arts and Culture 2003 Conference Proceedings, Melbourne, May 2003

McGonigal, J.: Reality is Broken: Why Games Make us Better and How they Can Change the World. Jonathan Cape, London (2011)

Miller, H.K.: Review: exit through the gift shop. Sight Sound **20**(5), 65 (2010)

Mixed Reality Lab: Human PacMan. http://mixedrealitylab.org/projects/all-projects/human-pacman/ (no date). Accessed 1 Apr 2016

Montola, M., Stenros, J., Waern, A.: Pervasive Games: Theory and Design. Morgan Kaufmann, London (2009)

Murray, H.J.R.: A History of Board Games Other Than Chess. Clarendon Press, Oxford (1952)

Nevelsteen, K.J.L.: A Survey of Characteristic Engine Features for Technology-Sustained Pervasive Games. Springer, London (2015)

Nijholt, A.: Human avatars in playful and humorous environments. 2016 AHFE International Conference: 27–31 July 2016 Walt Disney World, Swan and Dolphin Hotel, Florida (Forthcoming)

Oaklandish: Oaklandish.com http://oaklandish.com/about (2016). Accessed 1 Apr 2015

Oldenburg, R.: The Great Good Place: Cafes, Coffee Shops, Community Centers, Beauty Parlors General Stores, Bars, Hangouts and How They Get You Through the Day. Paragon House. Paragon House, New York (1989)

Ordnance Survey: Sheet 28: Dartmoor. B3 edition. 1:25,000. OS Explorer Map. Ordnance Survey, Southampton (2010)

Pac Manhattan: New York University Game Center. http://gamecenter.nyu.edu/pac-manhattan/ (2014). Accessed 10 Mar 2015

PacManhattan: Come Out and Play Festival. http://www.comeoutandplay.org/project/pacmanh attan/ (2014). Accessed 10 Mar 2015

Palmer, J.: Let's Go Letterboxing: A Beginner's Guide. Orchard Publications, Newton Abbot (1998)

Parlett, D.: The Oxford History of Board Games. Oxford University Press, Oxford (1999)

Posterchild: How to make your own totally sweet Mario Question Blocks and put them up around town because it's really awesome. Qwantz.com [Internet Archive] http://web.archive.org/web/20050609234453/ http://www.qwantz.com/posterchild/ (2006). Accessed 10 Mar 2015

REXplorer: Museum Mobile Wiki. http://wiki.museummobile.info/museums-to-go/products-servi ces/rexplorer (no date). Accessed 12 Mar 2015

REXplorer: Youtube. (no date). Accessed 12 Mar 2015

Salen, K., Zimmerman, E.: Rules of Play: Game Design Fundamentals. MIT Press, Cambridge (2004)

Suits, B.: The Grashopper: Games, Life and Utopia. Broadview Press, Peterborough, Ontario (2014)

Thacher, S., Findley, U.: The Fool's Guide to Immersive Narrative Adventure. IndieCade 2013. Youtube. https://www.youtube.com/watch?v=qWUmNGqY_7w (2013). Accessed 20 Mar 2015

Thomas, B., et al.: ARQuake: An outdoor/indoor augmented reality first-person application. In: ISWC2000: Proceedings of the Fourth International Symposium on Wearable Computers. October 2000, Atlanta Georgia USA. IEEE. http://www.tinmith.net/papers/thomas-iswc-2000.pdf (2000). Accessed 1 Apr 2016

Walker, A.: Here's Why the Banksy movie Is a Banksy prank. Fast Company, April 15. http://www.fastcompany.com/1616365/heres-why-banksy-movie-banksy-prank (2010). Accessed 25 May 2016

Wikipedia: Assassin (game). Last modified February 27. https://en.wikipedia.org/wiki/Assassin_(game) (2006). Accessed 20 Mar 2015

Woo, S.: Urban scavenger hunt finds followers. The Wall Street Journal, March 24. http://www.wsj.com/articles/SB10001424052748703858404576214842085391356 (2011). Accessed 25 May 2016

Zimmerman, E.: Jerked around by the magic circle: clearing the air ten years later. Gamasutra. http://www.gamasutra.com/view/feature/6696/jerked_around_by_the_magic_circle_.php (2012). Accessed 20 Mar 2015

Filmography/Videography

Artificial Intelligence: A. I. Directed by Steven Spielberg. Warner Brothers/Dreamworks SKG (2001)

Exit Through the Gift Shop: Directed by Banksy. Paranoid Pictures (2010)

Ikea Heights: Directed by David Seger. Channel 101 (2009–2010)

It's a Mad, Mad, Mad, Mad World: Directed by Stanley Kramer. United Artists (1963)

La Decima Vittima [The Tenth Victim]: Directed by Elio Petri. Compagnia Cinematografica Champion/Les FIlms Concordia (1965)

The Institute: Directed by Spencer McCall. Argot Pictures (2013)

Mapping the Beach Beneath the Street: Digital Cartography for the Playable City

Paul Coulton, Jonny Huck, Adrian Gradinar and Lara Salinas

Abstract Maps are an important component within many of the playful and gameful experiences designed to turn cities into a playable infrastructures. They take advantage of the fact that the technologies used for obtaining accurate spatial information, such as GPS receivers and magnetometers (digital compasses), are now so widespread that they are considered as 'standard' sensors on mobile phones, which are themselves ubiquitous. Interactive digital maps, therefore, are widely used by the general public for a variety of purposes. However, despite the rich design history of cartography digital maps typically exhibit a dominant aesthetic that has been designed to serve the usability and utility requirements of turn-by-turn urban navigation, which is itself driven by the proliferation of in-car and personal navigation services. The navigation aesthetic is now widespread across almost all spatial applications, even where a bespoke cartographic product would be better suited. In this chapter we seek to challenge this by exploring novel neocartographic approaches to making maps for use within playful and gameful experiences designed for the cities. We will examine the potential of design approaches that can produce not only more aesthetically pleasing maps, but also offer the potential for influencing user behaviour, which can be used to promote emotional engagement and exploration in playable city experiences.

P. Coulton (✉) · A. Gradinar · L. Salinas
Imagination, Lancaster Institute for the Contemporary Arts, The LICA Building,
Lancaster University, Bailrigg, Lancaster LA1 4YW, UK
e-mail: p.coulton@lancaster.ac.uk

A. Gradinar
e-mail: a.gradinar@lancaster.ac.uk

L. Salinas
e-mail: l.salinas@lancaster.ac.uk

J. Huck
School of Environment, Education and Development, Arthur Lewis Building,
The University of Manchester, Manchester M13 9PL, UK
e-mail: jonathan.huck@manchester.ac.uk

© Springer Science+Business Media Singapore 2017
A. Nijholt (ed.), *Playable Cities*, Gaming Media and Social Effects,
DOI 10.1007/978-981-10-1962-3_7

Keywords Digital cartography · Neocartography · Digital maps · Game design · Location-based games

1 Introduction

The title of this chapter includes the phrase "*The Beach Beneath the Street*", which refers to the graffiti "*Sous les pavés, la plage!*" that appeared in Paris during the civil unrest of May 1968. This phrase is often translated as "*Beneath the pavement—the beach!*" (Wark 2015) and is most often associated with the influential French Avant Garde group the Situationist International (SI) who took part in the street demonstrations. The SI argued, in a similar vein to Lefebvre (1991), for a right to view the city as a space for democratic possibilities, a social geography of freedom within which the rules of everyday life would be turned upside down and restored into a "*realm for play*" (Shepard and Smithsimon 2011). One of the key concepts of the SI is '*Detournement*' which is French for '*rerouting*' or 'hijacking' and was used to describe the reuse of pre-existing elements to create a new ensemble (Debord and Wolman 1981). One of the main SI practices is the *dérive* (drifting), a technique of rapid passage through varied ambiances. Dérives involve playful-constructive behaviour and awareness of psychogeographical effects, and is different from the classic notions of journey or stroll.

Psychogeography was first introduced into the SI discourse by Guy Debord in his essay Introduction to a Critique of Urban Cartography:

> Psychogeography could set for itself the study of the precise laws and specific effects of the geographical environment, whether consciously organized or not, on the emotions and behavior of individuals. The charmingly vague adjective psychogeographical can be applied to the findings arrived at by this type of investigation, to their influence on human feelings, and more generally to any situation or conduct that seems to reflect the same spirit of discovery.

The SI celebration of revolutionary play is revived in creative explorations of location-based technologies. These location-based practices often adopt the 'tracing while wandering' tactic of the dérive to map the city's geography (Tuters and Varnelis 2006). From media art to human computer interaction design, these situated practices 'bridge the digital with the analog' to 're-imagine urban space from the point of view human subjectivity' (Tuters 2011). The concept of a 'Playable City' echoes the desire of re-appropriating and re-mapping our urban infrastructures as spaces of play and provides a useful lens through which we can consider the games and playful experiences discussed in this chapter.

Whilst there is a long history of street games (Culin 1891) that could be explored within this chapter, in this research we concentrate on the games that have emerged through the utilisation of mobile phones and their infrastructures. In particular, we will focus upon the so-called 'Location Based Games' (LBG), in which the gameplay evolves and progresses in relation to the player's physical location. Inherent within this definition is a requirement for a player to be able to indicate their location during gameplay. However, whilst providing location is increasingly abstracted from the developer at a technical level, it remains a complex process (Rashid et al. 2006a, b) and a wide variety of techniques have emerged. These techniques take advantage of either the mobile infrastructure, for example the cell id, or sensing technologies embedded into mobile phones such as GPS receivers. Perhaps unsurprisingly, the earliest LBG such as Botfighters, Undercover, The Journey, and I Like Frank all used cell id (Rashid et al. 2006a, b) although its prevalence was relatively short-lived due to the emerging availability of GPS, initially through separate devices connected via Bluetooth and later through GPS chipsets being built into mobile phones. The most well known of the early GPS games is arguably Mogi from Newt Games, which achieved notable success in Japan (Licoppe and Inada 2008). Other early games experimented with implied location solutions where players interacted with objects or systems that have a known location using such things as NFC (Coulton et al. 2006), Bluetooth beacons (Rashid et al. 2005), and visual markers (Chehimi et al. 2008).

Whilst all of these early games used location, they preceded the availability of mobile mapping services, and so used either bespoke maps of a specific area or adopted a radar style visualisation by simply showing relative proximity to other game players or objects. Further, the resolution of these visualisations was severely restricted by the resolution of the mobile screens as illustrated in Fig. 1 which shows the map for the location-based game PAC-LAN (Rashid et al. 2006a, b). This map represents a physical game space of approximately 0.5 km^2 on a 128 × 128 pixel screen.

Fig. 1 First PAC-LAN game map (*left*) and the area it represents (*right*)

Maps constitute a ubiquitous medium through which we understand, construct and navigate our surroundings (Ballatore and Bertolotto 2015). As interactive digital maps have increasingly replaced paper maps, this *"ubiquitous cartography"* is quietly becoming part of our daily lives (Ballatore and Bertolotto 2015). One of the major drivers for this was the launch of the Google Maps API in June 2005, initiating a wave of location-based services and mash-ups. The release of Google Maps for Mobile in 2007 and its API in 2008 initiated a further developer frenzy and whilst other mapping services have appeared and subsequently disappeared, it undoubtedly remains the dominant mapping service. It is widely acknowledged amongst cartographers and map-makers that the advent of 'mobile maps' has had a profound impact upon the way in which we interact with mapping in general. *"Map Readers"* have become *"Map Users"* (Muehlenhaus 2013), and their expectations have changed accordingly. Users no longer look at maps to find where they are, but rather they 'tell' the map where they are and expect the map to orientate itself around them (Clark 2008). The generic nature of these maps, however, means that finding the geographical information relevant to the task at hand can often be a challenge for users, as they are faced with a 'default' map containing standard feature content that does not account for their specific needs or preferences (Wilson et al. 2010). In an attempt to combat this, map providers typically overlay their maps with information (often described as Points of Interest (POI)) that may be of some relevance to the user, based upon stored information about them. Unfortunately, some of this information will naturally be superfluous, and can lead to "spatial information overload", whereby irrelevant feature content can actually hinder users, and in particular, task goals (Wilson et al. 2010).

Whilst tools such as Google Maps make it 'simple' to create digital maps, they provide very little scope for aesthetic customization beyond simply re-colouring map features. The market dominance of these maps has led to expectations amongst map users that a digital map will look and behave like a Google Map. A further expectation is that can be quickly produced and without the requirement for 'expert' decision making with regards to symbology, projection, scale and so on (Muehlenhaus 2013). It is not surprising, therefore, that many cartographers have lamented the rise of these commercial mapping tools, which come with predefined projection, scale, typography, colour and symbology (Field 2013). It is argued that such predefined digital maps are creating a global *"blandscape"* and a sense of *"unauthordness"* through the lack of detail and apparent "homogenization" of the landscape (Kent 2008). This *unauthordness*, also evident in LBG, is disappointing given that other types of game have a strong history of making extensive use of maps. The purpose of this chapter is to consider why this *blandscape* has been created, and how we might encourage designers that cartography can take a greater role in the creation of playable cities.

2 Cartography–Neocartography

It has been widely reported that the discipline of cartography has "*struggled to keep up*" with rapid developments in digital mapping technologies over the past quarter-century (Muehlenhaus 2013), with each major technological development in map-making (desktop GIS, web mapping, mobile mapping) being lauded as the "*death of cartography*" by many commentators (e.g. Wood 2003; BBC 2008). Indeed, it is claimed that many modern maps, particularly those on the web, are now produced "*without cartography*", as the projection, scale, typography, colour and symbology are all predefined by the map provider (Field 2013). In an age where cartographic styles may be considered more as a commercial asset for brand recognition than a device for communication, there is little question that the craft of online map-making has developed without too much input from the discipline of cartography itself (Muehlenhaus 2013).

It should be noted, however, that from the advent of thematic cartography in the early 1800s right up until the 1930s, mapping would be undertaken by architects, engineers and designers; not those specifically trained in map design (Muehlenhaus 2013). Prior to this, maps were purely objects of reference, and only used for demarcating property ownership, state territory, urban layouts and war (Pickles 2004). The word '*cartography*' is, in fact, a relatively new term: it was coined as a Portuguese neologism in 1839, and only appeared in the Oxford Dictionary in 1859 (Wood 2003). Cartography only developed as a field of academic interest following the Second World War, during which it was realised that maps were not only useful tools of objective visualisation, but also powerful rhetorical tools that could be used, for example, to spread propaganda (McMaster and McMaster 2002; Muehlenhaus 2013). It may be considered, therefore, that maps have in fact been produced "*without cartography*" for the vast majority of their history, with only a brief period between the 1930s and the 1980s where institutionalised cartographic conventions prevailed. This view is supported by Wood (2003), who describes the conflation of map-making and cartography as "*at best anachronistic, at worst unpardonably presumptuous*". Modern developments in the field of digital neocartography may also be separated from the '*traditional*' body of academic cartography, with a focus that is much more seated in the practices of design than cartography (Wood 2003).

Muehlenhaus (2013) questions whether Google Map has "*revolutionised*" or "*euthanized*" map-making, with the ubiquity of Google Maps now effectively meaning that people are coming to expect that all maps should look and behave like Google Maps. The result of this is, of course, that many more maps are created using services from commercial mapping providers such as Google. In turn, this has further reduced the demand for designed-for-purpose cartographic products, leading to the subsequent reduction in the development of cartographic tools and software, arising from insufficient demand. In their place we have seem a practice emerge in which the regular creation of maps is performed by almost anyone, with little regard for data, design philosophies, or principles of mapping (Field 2013). Furthermore, the ability to make such maps more easily has, in many cases, led to the expectation that should be made quickly, once again at the expense of the rigorous design process required for a high-quality cartographic product (Field 2013). This results in an over-reliance upon "*dumb defaults*", and the impression that the "*art*" of map-making is being lost as all maps start to look the same (Field 2013).

Nevertheless, mapping, like writing, can be done lucidly and elegantly (Monmonier 1993), and there is little doubt that recent developments in the field of digital cartography have permitted the development of approaches to mapping that are able to communicate information more effectively, and even have the potential to affect the behaviour of the map-reader (Huck et al. 2015, 2016; Gradinar et al. 2016). Irrespective of the specific purpose of a map, the era within which it was created, or the media or technology with which it is produced, the primary goal remains the same: a map seeks to communicate some form of information (Field 2013; Muehlenhaus 2013); and for this to be achieved, modern digital cartography must strike a balance between art, science and technology in order to be effective (Field and Demaj 2012). A beautiful map based upon miscalculated, poorly measured or otherwise incorrect data may be considered to be of little utility; and conversely an ugly or unclear map, however accurate, is unlikely to inspire the confidence of the map-reader, also limiting its utility (Wright 1942).

In recent years, the reported shift towards the paradigm of neogeography (Turner 2006) has been accompanied by much discussion about the emergence of neocartography (Kraak 2011). Whilst the definitions for terms such as these can be somewhat nebulous, Kraak (2011) describes a "*working definition*" for neocartography as "*a limited set of loosely coupled simple guidelines that allow an adequate visualization of large amounts of diverse data sets*". For the purposes of this chapter, neocartography may be considered as the area of cartography that deals with the automated rendering of multi-scale data using style sheets or similar rule-based approaches, with little or no manual intervention from the cartographer or designer. This contrasts with the traditional approaches to cartography, in which the cartographer or designer would interact manually with many map elements, in order to produce a map that would normally represent a defined region at a given scale. Neocartography, therefore, may certainly be considered to encompass those technologies that produce tiles of data at multiple scales to populate the 'slippy map' interfaces of neogeographical web maps such as Google Maps, Bing Maps and OpenStreetMap. Data sources such as OpenStreetMap, technologies such as Mapnik and companies such as Stamen Design and MapBox have been instrumental in bringing neocartography into popular use, through the release of open-data and open-source software that allows individuals to create and implement their own neocartographic style sheets. These developments have, in turn, allowed the creation of exciting new possibilities in the production of novel digital mapping content (Huck et al. 2015).

3 Mapping in Games

Arguably, the history of games is as old as the history of maps, with evidence of board games being played for more than 5000 years (Whitehill 1999). Maps have played an important role in many games throughout history: treasure maps in board games, schematic reality-inspired maps in Monopoly, maps of fantasy worlds in video games, and so on. Perhaps the most well known, however, is Chess, which, like Draughts and Go, was once a real-world simulation in which players assume the role of an army leader charged with the objective to overcome an enemy and conquer territory (Ahlqvist 2011). The primary role of maps in games is to support the game mechanics and rules. Many of the early games employed very simple and abstract maps, replacing 'realistic' looking cartography with very basic representations of territory that were much easier to replicate (e.g. the chequered pattern on a chess-board). This changed in the mid-nineteenth century when advances in lithography and production techniques allowed games to be printed in large commercial quantities and coincided with an expansion of games away from war-like simulations to encompass a wide variety of narratives (Ahlqvist 2011). Whilst the history of maps in physical games deserves a more detailed discussion, in this chapter we are concerned with computer games and so such a discussion is, unfortunately, beyond its scope.

In the case of early computer games, it is often tempting to begin the discussion with arcade and console games such as Pong, Space Invaders, Lunar Lander, and so on. The advent of maps in computer games predate these with the 1975 text-based game adventure and the first multiuser dungeons (MUDs) invented by Roy Trubshaw and Richard Bartle in 1978, which are arguably the first significant developments of mapping in computer games. The main component of these first MUDs was a database that was used to define the rooms, objects, commands, etc., which was defined in a separate file from the main game code. This meant that players were able to add new features, or otherwise alter the game, by editing these files. Despite the games providing either no visual information or graphics based upon ASCII characters (as shown in Fig. 2), the games themselves, along with the players' modifications, perhaps best represent the first examples of cartography emerging within computer games. This practice of altering games became known as 'modding', and persisted as maps continued to became integral within computer games. This was particularly clear with the emergence of first-person shooters (FPS) such as Doom in 1993, which allowed third parties to create custom levels, effectively designing their own maps. Modding became such a popular activity that level editors appeared in early 1994 and indeed the practice became the entry point for a number of future game designers (Sotamaa 2010).

Mapping has seen a resurgence in the most recent generations of video games, which have been described as "*the bold future of cartography*" (Garfield 2012), and have often been cited as a notable exception to the perceived stagnation of map production. Games such as '*Grand Theft Auto V*' by Rockstar Games (2014) and '*Elder Scrolls V: Skyrim*' by Bethesda Game Studios (2011) comprise not only

Fig. 2 Example of an ASCII MUD map

detailed three-dimensional '*open world*' maps to explore within the game, but also 2D on-screen maps for navigation, and even accompanying paper maps. '*Skyrim*', for example, comes with a fold-out stylised map on textured faux parchment and a game guide that includes 220 pages of maps, representing the work of many digital cartographers (Garfield 2012). Further, cartography still dominates much of the current modding activities with numerous sites offering maps for popular games such as *Half Life*, *Counterstrike* and *Minecraft*. Given this interest, maps can be seen as a vital component of modern game production. Having discussed maps in games more generally, we will now consider the role of maps in games designed to facilitate a playable city.

3.1 Mapping in Location-Based Games

The concept of 'location based' game came to the fore with the removal of 'Selective Availability' from the GPS network in May 2000. The resulting proliferation of hand-held GPS receivers meant that, for the first time, the general public had access to their own location via the GPS network. Geocaching, arguably the first digital location-based game, was first played that same month, with players initially using GPS-derived coordinates to hide and seek '*caches*' (reminiscent of the traditional orienteering game letterboxing). The coordinates of caches were provided in online lists, and the 'caches' often took the form of a waterproof container containing a logbook as well as some trinkets which players are encouraged to trade through the caches. As GPS receivers started to be embedded into mobile phones, a dedicated application for Geocaching was created that included a Google Map with icons to assist players in finding '*caches*'. Nevertheless, similar games such as Munzee have been much faster on the uptake of emerging smartphone technologies, including QR codes and NFC tags, and as a result are attracting many former Geocaching players. Today, there are more than 2.5 million geocaches in over 180 countries and more than 10 million registered users on Geocaching.com.

As already identified, the map component of the Geocaching mobile app is a Google Map. Whilst, prior to the availability of commercial web mapping services, some games created their own customised maps (e.g. Rashid et al. 2006a, b), Google Maps now dominate in location-based gaming. For example, the mafia style type of game TurfWars[1]; the role-playing games such as Parallel Kingdom[2] and Global Supremacy[3]; and the academic research game Urbanopoly (Celino 2012) all rely upon Google Maps. The reason for this is clear: Google Maps provide a familiarity in appearance and function that has come to be expected by users, and an ease of implementation that is attractive to developers. Nevertheless, there are some examples where bespoke maps have been implemented within LBG. The now discontinued Shadow Cities[4] is one such example, as is the augmented reality game Ingress,[5] which makes use of the Google Maps data to create a personalised map for each player and was subsequently used to devlop PokemonGo[6]. However, such games are few and far between and, whilst their bespoke maps do contribute positively to the game aesthetic, they make no attempt to explore the impact that novel approaches to map design could have upon the gameplay.

Whilst not LBG per se, activity fitness tracking applications are increasingly making use of digital maps. Once again, the dominance of commercial mapping providers such as Google Maps is demonstrated by applications such as Strava,[7] Endomondo,[8] Runtastic[9] and Nike Plus.[10] To enhance the user's immersion levels, these applications have been augmented with additional content. Examples of this include Runtastic, which includes a "*Story Running*" activity,[11] where users can

---

[1]https://turfwarsapp.com.

[2]www.parallelkingdom.com.

[3]www.globalsupremacyapp.com.

[4]www.engadget.com/tag/Shadow-Cities.

[5]www.ingress.com.

[6]www.pokemongo.com

[7]www.strava.com.

[8]www.endomondo.com.

[9]www.runtastic.com/en/apps/runtastic.

[10]www.nike.com/gb/en_gb/c/running/nikeplus/gps-app.

[11]www.runtastic.com/en/storyrunning.

listen to a diverse range of stories intended to enhance the running experience; and Zombies Run,[12] which adds a narrative into '*missions*', enhancing the activity with games like adversaries such as zombies 'chasing' the player, or the achievement of in-game quests that are required to '*build your base*' and '*defend your community*'. Finally other applications, such as Swarm (formerly Foursquare), may be called 'quasi-games' because they do not have very well-established rules so may be considered as occupying a 'middle-ground' somewhere between social networking tools and games (Lammes 2011). Whilst such quasi-games exhibit play-like elements, their status as games is not universally accepted (Deterding et al. 2011) and the majority of these utilise the same navigational aesthetic.

Having highlighted the importance of maps in games and a requirement for greater consideration to the design of maps in LBG and playful experiences, we now consider how this can be practically achieved.

4 Designing Maps for Playable Cities

We believe that digital map aesthetics can be optimised for a variety of given purposes and in this section we present two case studies that explore this through the novel cartographic techniques of '*feature abstraction*' and '*dynamic visual hierarchies*' in order to demonstrate how aesthetic decisions can alter the experience and even behaviour of users (Gradinar et al. 2016).

4.1 PAC-LAN Zombie Apocalypse: Feature Abstraction

The original PAC-LAN location-based game was created in 2006 to consider the possibilities arising from the emerging combination of mobile phones with near-field communications (NFC), at the time using the general name of radio-frequency identification (RFID), to create new location-based experiences (Rashid et al. 2006a). The game was a novel implementation of the arcade classic Pac-Man, which was chosen as it allowed a comparison with other location-based versions of the game developed around the same time, such as '*Pac Manhattan*' and '*Human Pac-Man*' (Rashid et al. 2006a). Principally, the game investigated using NFC as an alternative to GPS as the method of determining location by requiring players to tag physical objects at known locations.

As the game proved successful and technology has matured considerably in the intervening period, a new version of the game was developed in 2015 with the focus much more on the overall experience rather than the interaction with the technology (Gradinar et al. 2015). PAC-LAN: Zombie Apocalypse was built using the Android platform to allow the game to be distributed '*in the wild*', as Android supports the greatest number of handsets with on-board NFC readers. Similar to the original

---

[12]https://zombiesrungame.com.

PAC-LAN game, the *'pills'* that Pac-Man collects during the original arcade game were recreated as physical items using yellow and red Frisbees. The pills also included straps so they could be attached to the physical environment (e.g. lamp post, drainpipes, trees, etc.) and an NFC tag, allowing access to the name and location of that pill via the game system.

The game supports five players, each of which adopts a particular character persona (indicated by wearing a character hat) and are identified by a NFC tag along with an Android mobile device running the game application. One player takes the role of PAC-LAN, whose purpose is to tag all of the pills in order to earn points; and the remaining four players take the role of the Zombies, whose purpose is to *'infect'* the PAC-LAN player by scanning their NFC character tag. As with the arcade version, there are a number of special 'power pills' located within the game arena which, when tagged by PAC-LAN, enable the player to *'neutralise'* the Zombies and forcing them to return to the starting location of the game to *'re-spawn'*. A key aspect of the game design is that all of the main interactions within the game (such as tagging pills or characters, for example) occur physically and do not require any direct interaction with the interface of the mobile screen.

Before starting a particular game of PAC-LAN the game space needs to be set up, which involves positioning the pills both physically in the landscape and digitally on the map, which is undertaken within the mobile application. There are three types of pills in the game:

1. The base pill: This is the location where all players start the game and where players 're-spawn' after being neutralised (Zombies only) or if their *'roaming time'* expires (all players).
2. Check-in pills: These pills update the last tagged location of all active players in the game database, as well as replenishing their 'roaming time' when they tag it, allowing them to stay in the game for longer.
3. Power pills: These pills give PAC-LAN the ability to *'neutralise'* Zombies whilst preventing the Zombies from being able to *'infect'* PAC-LAN. They also replenish the *'roaming time'* and extra time for 'infecting' zombies. These pills are only distinguished on the PAC-LAN's UI as for Zombies they function the same as check-in pills.

Once the physical game space has been set up, each player tags the base pill to start their game. The PAC-LAN player begins play immediately (Fig. 3 left), whereas each Zombie player has a predefined waiting time of between 100 and 40s that must expire before they can join the game (Fig. 3). This gives the PAC-LAN player time to safely exit the immediate vicinity of the Zombies before they can give chase.

The game user interface (UI) shown in Fig. 4 consists of a map (considered in detail in forthcoming section) depicting the location of all pills, paths and buildings, and a decreasing progress bar (*'roaming time'*) that represents the maximum time that a player can take between tagging pills before being forced to return to the base. Whilst returning to the base, the player is inactive and so no interactions with the UI or game pills are allowed until *'the base pill'* is scanned again. When a player tags a pill, the *'roaming bar'* is replenished and the UI refreshed by showing the location

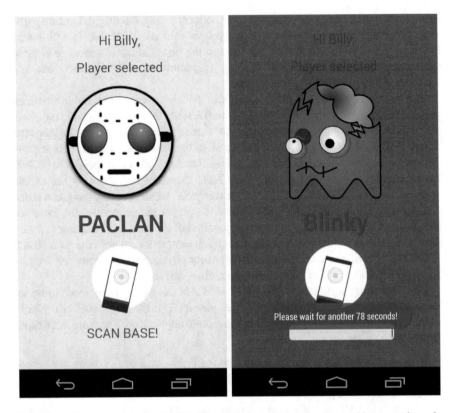

Fig. 3 PAC-LAN entering the game (*left*) and Zombie entering the game with a countdown for 'waiting time' (*right*)

of the last pill tagged by each active player, as well as changing the colour of the pill to grey to indicate it is no longer worth any points. Each player's interaction with the pills is visible only on their individual UIs. The player can return and '*re-tag*' a grey pill to replenish their '*roaming time*' and to update the last tagged location of all other active players, though they will not receive any more points for doing so.

Whilst the previously described features change the sequel (2015) considerably from the original PAC-LAN game (2006), it is the digital game map that is the primary differentiator. In particular, the introduction of the smartphone instead of the original feature phone permitted the design of a high-resolution, dynamic '*slippy*' map that could be generated in real time based upon real geographical data, as opposed to the low-resolution static map that had been used in the previous version.

In the design of the map for this game, three design goals were considered:

1. Promote immersion into the game through the use of a suitable aesthetic.
2. Perform well in the context of a mixed reality LBG (running, outdoors).
3. Encourage players to navigate '*head-up*' rather than '*head-down*'.

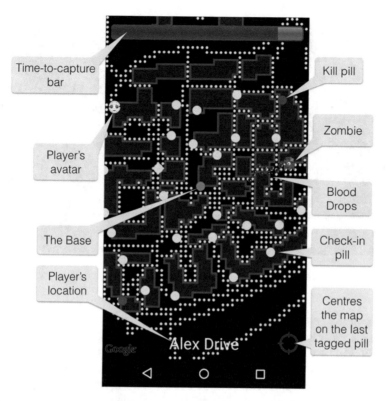

Fig. 4 PAC-LAN game user interface

In particular, the third design goal was of key importance to this game, as the research goal was to explore the Dichotomy of Immersion (Huck et al. 2015): that peculiar situation created in LBG's whereby a player's attention is constantly divided between the physical world within which the game is being played (the physical world), and the screen of their mobile device (the digital world). The split of attention between the physical and digital game components is usually dominated by inter-action with the screen at the expense of interaction with the landscape, which can limit emotional engagement with (and therefore immersion into) the LBG (Lei and Coulton 2011).

This design goal was approached by the use of '*feature abstraction*' as a carto-graphic device. It was hypothesised that a small amount of abstraction in map features may encourage players to '*look up*' more and verify what they see on the map against their physical surroundings, thus increasing their engagement with their physical surroundings. This is in contrast with the use of a more traditional (precise) map, which will not require validation against the landscape; or a map that is '*too abstract*', which may be too difficult to read quickly whilst playing, thus increasing interaction with the screen at the expense of the landscape.

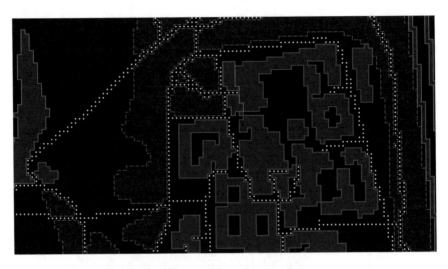

Fig. 5 The 'PAC-MAP' design for PAC-LAN

Four of the cartographic solutions to the above design goals will be presented. The maps have been all created using data derived from OpenStreetMap using PostGIS, and have been rendered using Mapnik. For the purpose of comparison, all maps are shown at the same standard orientation (north at the top), zoom level (17, equivalent to a geographic scale of 1:4514) and extent (showing part of the Lancaster University campus). None of the figures includes a legend, north arrow, scale bar or similar, as such features are not typically provided in LBG's. In-keeping with the above design goals, all of the maps are highly limited in terms of the number of features they contain, with only four or five feature classes typically included to facilitate easy reading. For the same reason, augmentations such as labels and POI have also been omitted from all of the map designs.

The first map presented is the Pac map (Fig. 5), which is designed primarily to match the aesthetic of the classic Pac-Man arcade game developed by Namco in 1980. Abstract feature representation has been achieved by generalising all nodes to the nearest 10 m, and forcing vertices to be oriented either north-south or east-west: resulting in abstract features that also contribute to the game aesthetic. There are only four feature classes included in the map: '*building*', '*path*', '*trees*' and '*hazard*', all of which are rendered using styles directly inspired by the Pac-Man game. Trees and buildings are drawn in the same blue as the Pac-Man maze, and a complementary red has been used to mark out hazards. In order to ensure that the dark palette employed by this map performs well outdoors, the lines in the map are thick, with very fine white lines drawn into the blue and red in order to increase their contrast with the black background. Pathways (including roads and footpaths) have been marked out with white dots, once again to gain contrast with the dark background, whilst also reflecting the 'pills' that Pac-Man collects from within the maze in the original game.

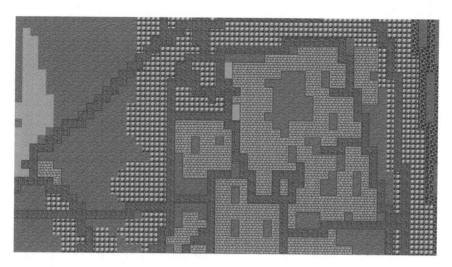

Fig. 6 The 'RPG Map' design for PAC-LAN

The second, RPG map (Fig. 6), is inspired by the '*classic*' role-playing games (RPGs) of the 1980s and 1990s. The data has been abstracted into a grid of 20-m cells, each of which can only contain one of five feature classes: '*building*', '*road*', '*water*', '*trees*' or '*hazard*'. Cells were then dissolved into contiguous areas of each data type, and coloured using tiled textures collected from freely available online sources. The use of a coarse 20-m grid gives this map a greater level of abstract feature representation than the Pac map, therefore making it more difficult to rely upon for navigation, in order to investigate the effect that this has upon the player's interactions during gameplay. The coarse grid, RPG-style textures and playful features (e.g. the use of a '*lava*' texture to denote hazards) lend a definite '*game aesthetic*' to the map, but in less-specific manner to the Pac map, permitting exploration as to the effect of this upon players' perceived level of immersion.

Third, the 'Sketchy Map' (Fig. 7) employs '*sketchiness*' as an approach to abstract feature representation, acting to obscure the precise position and shape of geographic features. The '*hand-drawn*' or '*sketchy*' effect on the polygons has been achieved by a combination of polygon smoothing, line smoothing, multiple-overlay and image composite operations in order to give the impression that they have been drawn using felt-tip pens (akin to the approach first suggested by Ashton 2012). Conversely, the line features were simplified using the Visvalingam–Whyatt line generalisation algorithm (Visvalingam and Whyatt 1993), and overlaid using transparency and image composite operations in order to give the appearance of having been drawn using highlighter pens. The main difference with this approach is that the level of abstraction varies from feature to feature as opposed to being uniform across the dataset as is the case in the grid-based approaches, which may prove more disorientating for users. The '*hand-drawn*' aesthetic promotes a '*playful*' feel to the map, but without specifically portraying a '*game*'.

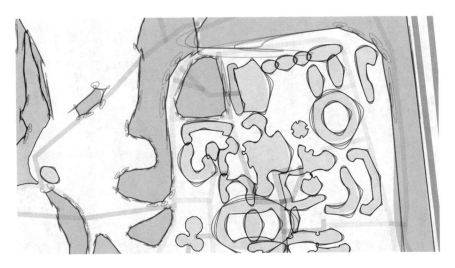

Fig. 7 'Sketchy Map' design for PAC-LAN

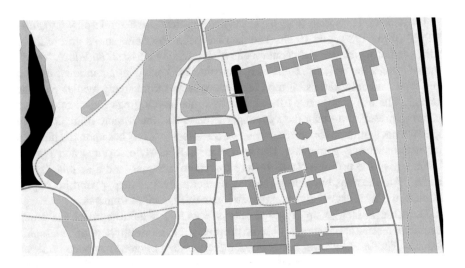

Fig. 8 The 'Anti-Glare Map' design for PAC-LAN

Finally, the '*Anti-Glare Map*', shown in Fig. 8, was created which includes no spatial ambiguity whatsoever, and instead is characterised by a triadic colour scheme in order to maximise the contrast between all map features. This map is designed as a benchmark from which players could evaluate the other three maps.

These maps were tested with eight players that were asked to play a short custom-built LBG that required them to navigate between 20 physical objects (the PAC-LAN game pills) with five objects per map type. The initial map was chosen at

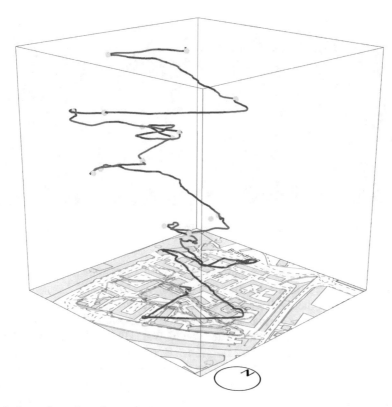

Fig. 9 Space time prism of map trial instance

random and then changed for another randomly selected map after every fifth pill
was correctly tagged, in order to allow the assessment of each map. The LBG tracked
players' progress using GPS and required them to hold down the volume rocker of
the mobile phone in order to view the map. This mechanic also means that players'
interactions with the map were logged alongside GPS-derived location for later
analysis using space-time prisms (Coulton et al. 2008) as shown in Fig. 9. In this
figure the base map is the physical area and time is represented vertically with the
location of the pills shown in yellow and blue and red indicating whether the map is
visible or invisible to the player, respectively. Following play, all eight players were
also given a semi-structured questionnaire and interview about how each map influenced their level of interaction with the screen and their surroundings.

The results of trialling the maps concluded that Pac map style map (Fig. 5) was
universally preferred by the participants and was also the most suitable of the maps
for the promotion of 'head-up' navigation in the game. The combination of the Pac-
Man gaming aesthetic and moderate feature abstraction promoted the game play
whilst encouraging the player to validate what they saw on the map against the
physical landscape. This result is interesting if viewed from a purely utility and

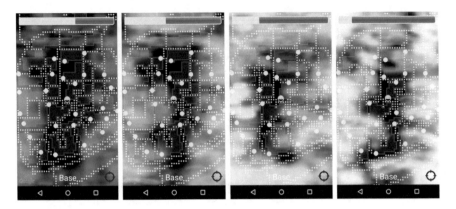

Fig. 10 Fog of War operating on four levels on the UI

efficiency perspective, which would clearly favour the anti-glare map and highlights the desirability for greater utilisation of cartography within LBG.

There are also other deliberate design features that obscure parts or the whole of the UI at different times during the game, in order to encourage the player to be less dependant upon the screen. First, The 'Fog of War' (FoW) shown in Fig. 10 is a game feature controlled by the '*roaming time*'. It consists of incrementally obscuring the map with a white '*fog*', thus making it more difficult to read. There are four different levels of '*fog*', with each level obscuring a greater area of the map as it incrementally '*creeps in*' from the edge of the screen towards the centre. To further aid the player's awareness of the passing of the 'roaming time', each level of the FoW is associated with a haptic feedback, with handset vibrations informing the player of their remaining '*roaming time*' without the requirement to check the screen. Second, '*Screen Blanking*' is designed to discourage over reliance of the player on the mobile screen. It consists of turning the whole mobile UI into a black screen after 5 s from the last action the user performed. There are two types of actions that make the screen fully visible again: either the player tags a pill or player, or explicitly requests access to visualise the game interface by tapping the mobile screen. The latter action displays the UI in its current form, and so does not remove the FoW. It is important to note that screen blanking and the FoW are not mutually exclusive.

An additional '*Blood Drops*' feature is used to assist the PAC-LAN player in finding Zombies whilst under the effect of a power pill, giving a slight advantage in their attempt to '*neutralise*' a Zombie. As soon as the PAC-LAN player tags a '*power pill*', a virtual trail of blood drops is displayed on PAC-LAN's map, which shows five GPS positions recorded after each Zombie last tagged a pill. Given that the PAC-LAN player does not know how long ago the Zombies last tagged a pill, this information indicates their initial direction of travel from their last tagged pill, but does not give away their location with any certainty. The game is finished when either PAC-LAN tags all the pills or is tagged by one of the Zombies and the winner is calculated based on the total points gained during the game. When the game finishes, the mobile application displays a leader board showing all five players and their scores.

Overall this study illustrates the benefits that designing maps with a particular aesthetics can have on playable city experiences and in the following we consider how hierarchy of the maps could also be used to change the nature of these experiences.

4.2 Paths of Desire: Dynamic Visual Hierarchies

Map design revolves around the need to satisfy a particular communication goal (Field 2013) relative to a given task. For commercial mapping services such as Google Maps, this goal is primarily to act as a road atlas. It is arguably inappropriate, therefore, that these maps are often used for a wide range of purposes without due attention being paid to the suitability of the map for the task at hand. For example, many traditional paper tourist maps represent selected POI around a city but frequently sacrifice scale and completeness in favour of adopting a visual style that matches the image of the city they are trying to portray (e.g. an historic city), and focus upon the features of the city which would be of interest to tourists (e.g. museums).

Previous research has demonstrated that map design can have a significant effect on route selection, and that the visual hierarchy of roads and footpaths is one of the primary influencers (Field and Demaj 2012). In this example, we present a novel Android application called 'Paths of Desire', which unlike traditional digital maps has a cartography that can be dynamically adjusted in real time, modifying the visual hierarchies of paths (including roads, footpaths, etc.) and POI in order to try to encourage visitors explore the city. It is intended that the varying visual hierarchies will encourage users away from main routes and motivate exploration of areas that would otherwise be ignored.

Experimentation using this design aim has focused upon the city of Lancaster in the United Kingdom, which is a small historic city in the North West of England. Lancaster is dominated by a main pedestrianised street that runs through the town centre and a circular trunk road that encloses the town centre. Both of these features significantly influence the flow of traffic and pedestrians, as they move through the city. These are shown in Fig. 11, along with the major POI and other often overlooked tourist attractions.

Lancaster City Council has been actively engaged in attempts to attract both residents and visitors to other parts of the city through outdoor art exhibitions, music festivals and the installation of new street furniture. Whilst these efforts have had some limited impact upon the problem, so far there have been no attempts to address the use of digital maps, which is the dominant way in which new visitors are likely to explore the city. The aim of this project is to consider alternative design solutions and determine the extent to which novel digital map design can be used to promote tourism in under-visited areas of the city.

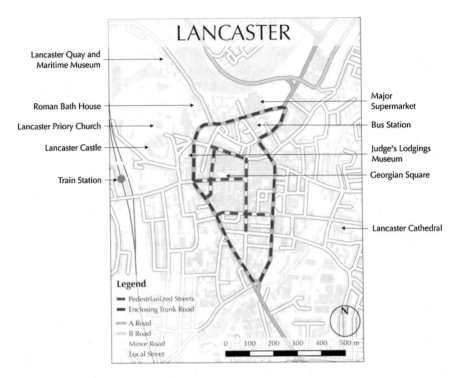

Fig. 11 A map of the city centre of Lancaster in the UK, illustrating the limited areas of the city commonly populated by visitors [Contains OS data © Crown copyright and database right (2015)]

To this end, an Android application has been developed to act as a 'digital tourist map' of the city. The design goal of the map is to promote exploration within the city using *'dynamic visual hierarchies'* to make features more or less attractive depending upon how many other tourists are currently located there (determined from the GPS location of other app users). The map design used in the application is defined by an XML style sheet that is downloaded from a server at launch. This allows for dynamic updates of the style sheet without the need to push updates of the application to users' devices. The initial map style guide was designed by the Manchester-based digital design agency Magnetic North, and is shown in Fig. 12.

The paths and POI are both supported by a three-level visual hierarchy based upon definitions under the control of the map designers. For example, this could be based on predefined values to promote particular routes and POI, or calculated in real time by monitoring the GPS-derived location of other app users in the city (i.e. with features classified as having the least visitors given hierarchical prominence). The screenshot on the left of Fig. 13 is an early attempt to recreate the dots of style sheet, which were not possible due to limitations in the vector tile rendering software upon which the application is built. As a result of this, the current version of the application uses three different line weights for the visual hierarchy, as is illustrated in the

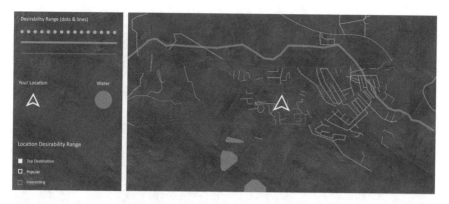

Fig. 12 Paths of desire style sheet developed by magnetic north

Fig. 13 Paths of desire applications screenshots

Fig. 14 Screenshot of the 'Lancaster Map' developed in collaboration with Chas Jacobs

screenshot to the right of Fig. 13. Ultimately as an extension to this research we intend to create a new vector tile map renderer that will allow us to fully realise the design and have greater flexibility in the developments of future designs than is currently possible.

Once the basic principles of dynamic visual hierarchies had been demonstrated using the application and map described above, we were able to start developing new maps for use in the application. It is intended that such maps may be easily created for each city that the application is used in, in order that the map may reflect the 'feel' of the city itself, as well as providing a pleasing aesthetic that would encourage potential users to look at and interact with it, and promoting playful exploration. An example of such a map is the 'Lancaster Map' that is shown in Fig. 14, which was developed in collaboration with the Lancaster-based artist Chas Jacobs.[13] Chas' art is well known around Lancaster, and often features the city itself, making him an ideal collaborator to achieve these design goals.

Using approaches similar to those described in the 'Sketchy Map' described above, the 'Lancaster Map' is designed to appear hand-drawn, with polygon smoothing techniques used in order to achieve this, and give the appearance of pencil guidelines around map features. The textures for the map were produced by Chas as

---

[13]www.chasjacobs.co.uk.

Fig. 15 A square of paint produced by Chas Jacobs, and the corresponding seamless texture tile used in the map

squares of paint on plain paper, which were then scanned into digital imaging software and processed into seamless tiles that are then applied to the smoothed shapes as textures. An example of a painted square and the corresponding seamless tile that is used for the 'woodland' texture is given in Fig. 15.

The maps themselves are implemented as a set of rule-based XML style sheets that may either be stored locally on the mobile device or downloaded from the server as required. The style sheets are then interpreted by the application renderer at runtime and styles are applied to features based upon the classification that is provided by the server. In this way, any number of optional maps may be implemented in the system for a variety of purposes, providing designs that suit the specific use-case of the application. For example, a map designed to guide tourists around historical attractions in a city might be antique in appearance, or one that it intended to facilitate playful exploration might look like a children's treasure map. In this way, the Paths of Desire application provides a novel approach to using cartography to affect behaviour, as well as a mechanism by with new map designs can be deployed quickly and easily on mobile devices.

The ability to dynamically change the design of maps as a reaction to external factors has a number of implications for the future of the role of maps. First, for the creation of the playable city in that dynamic in game events can be reflected directly upon the game map presented to the players, and second, it provides the ability to reconfigure the flow of people through a city in relation to particular requirements or events.

5   Discussion and Conclusions

Taking inspiration from the Situationists, our consideration of the playable city draws upon their notions of *detournement* ('rerouting' or 'hijacking') and *derive* (an awareness of psychogeographical effects) in order to remap the city for play. Drawing upon the rich history of games and maps, we have considered the playable city through the lens of neocartography. This approach recognises that whilst there can be little doubt that the widespread availability of digital and mobile maps has significantly impacted our interactions with and use of maps, these maps are being provided by a small number of commercial providers and exhibit limited opportunities for customization. This has resulted in the creation of a perceived '*blandscape*', in which the traditional purpose-driven cartographic designs have taken a back seat to the application of recognisable commercial colour schemes.

Whilst there have been calls for greater personalization of digital maps, the resulting solutions have been directed at usability, efficiency and clarity relative to the task being performed which has contributed to the homogenization of map aesthetics. It is unfortunate that this homogenization has also permeated through to the design of LBG which, in contrast with other types of computer game, have yet to realise the full potential of digital cartography.

In this research we have considered an alternative whereby the aesthetics of maps are used to affect the user experience in relation to exploration and serendipity within the playable city. First, we have presented location-based game that both adopt an aesthetic appropriate to the game but also use abstraction of features on the digital map to encourage '*head-up*' rather than '*head-down*' navigation thus promoting users' engagement with their physical surroundings. Second, we have presented a system capable of producing digital maps whose design can be adapted dynamically and in particular in which the hierarchy of POI, roads and paths can be adjusted in real time, with the aim of exploiting the known impact of map design upon behaviour in order to encourage users to explore their environment.

Overall, we believe that rich history of cartographic design has yet to be adequately addressed in relation to the aesthetics of digital maps, which we believe are an important component in realising the playable city.

Acknowledgments The research presented in this paper has been made possible through the Arts and Humanities Research Council (AHRC) project (AH/J005150/1), the Creative Exchange at Lancaster University in collaboration with Manchester design agency Magnetic North and the Lancaster-based artist Chas Jacobs.

References

Ahlqvist, O.: Converging themes in cartography and computer games. Cartogr. Geogr. Inf. Sci. **38**(3), 278–285 (2011)
Ashton, A.: Sketchy maps with geometry smoothing. https://www.mapbox.com/blog/sketchy-maps/ (2012). Last accessed 14 Mar 2016

Ballatore, A., Bertolotto, M.: Personalizing maps. Commun. ACM **58**(12), 68–74 (2015)

BBC News: Online maps 'wiping out history'. http://news.bbc.co.uk/1/hi/uk/7586789.stm (2008). Last accessed 14 Mar 2016

Celino, I., Cerizza, D., Contessa, S., Corubolo, M., Dell'Aglio, D., Valle, E.D., Fumeo, S.: Urbanopoly—a social and location-based game with a purpose to crowdsource your urban data. In: Privacy, Security, Risk and Trust (PASSAT), 2012 International Conference on and 2012 International Conference on Social Computing (SocialCom), pp. 910–913. IEEE (2012)

Chehimi, F., Coulton, P., Edwards, R.: Using a camera phone as a mixed-reality laser cannon. Intl. J. Comput. Games Technol. **2008**, 2 (2008)

Clark, J.: The new cartographers—what does it mean to map everything all the time? In these times http://www.inthesetimes.com/article/3524/the_new_cartographers (2008). Last accessed 14 Mar 2016

Coulton, P., Rashid, O., Bamford, W.: Experiencing 'touch' in mobile mixed reality games. In: Proceedings of International Conference in Computer Game Design and Technology (2006)

Coulton, P., Bamford, W., Cheverst, K., Rashid, O.: 3D Space-time visualization of player behaviour in pervasive location-based games. Intl. J. Comput. Games Technol. **2008**, 2 (2008)

Culin, S.: Street games of boys in Brooklyn, NY. J. Am. Folklore **4**(14), 221–237 (1891)

Debord, G., Wolman, G.J.: Situationist International Anthology (trans. Ken Knabb). Bureau of Public Secrets, Berkeley, California (1981)

Deterding, S., Sicart, M., Nacke, L., O'Hara, K., Dixon, D.: Gamification: using game-design elements in non-gaming contexts. In CHI'11 Extended Abstracts on Human Factors in Computing Systems, pp. 2425–2428. ACM (2011)

Field, K.: Editorial. Cartomyopic musings. Cartogr. J. **50**(1), 1–7 (2013)

Field, K., Demaj, D.: Reasserting design relevance in cartography: some concepts. Cartogr. J. **49**(1), 70–76 (2012)

Garfield, S.: On the Map: Why the World Looks the Way It Does. Profile, London (2012)

Gradinar, A., Huck, J., Coulton, P., Lochrie, M., Tsekleves, E.: Designing for the dichotomy of immersion in location based games. In: Proceedings of Foundations of Digital Games (2015)

Gradinar, A., Huck, J., Coulton, P., Salinas, L.: Beyond the blandscape: utilizing aesthetics in digital cartography. In Proceedings of the 2016 CHI Conference Extended Abstracts on Human Factors in Computing Systems, pp. 1383–1388 (2016)

Huck, J., Coulton, P., Gradinar, A., Whyatt, D.: Cartography, location-based gaming and the legibility of mixed reality spaces. In: Proceedings of 23rd GIS Research UK Conference (2015)

Huck, J., Gradinar, A., Coulton, P., Salinas, L.: Paths of desire: dynamic visual hierarchies to intentionally influence route decision. In: Proceedings of 24th GIS Research UK Conference (2016)

Kent, A.J.: Cartographic blandscapes and the new noise: finding the good view in a topographical mashup. Bull. Soc. Cartogr. **42**(1), 29–37 (2008)

Kraak, M.J.: Is there a need for neo-cartography? Cartogr. Geogr. Inf. Sci. **38**(2), 73–78 (2011)

Lammes, S.: The map as playground: location-based games as cartographical practices. In: Think, Design, Play, pp. 1–10 (2011)

Lefebvre, H.: The Production of Space, vol. 142. Blackwell, Oxford (1991)

Lei, Z., Coulton, P.: Using deliberate ambiguity of the information economy in the design of a mobile location based games. In Proceedings of the 15th International Academic MindTrek Conference: Envisioning Future Media Environments, pp. 33–36. ACM (2011). doi:10.1145/21 81037.2181044

Licoppe, C., Inada, Y.: Geolocalized technologies, location-aware communities, and personal territories: the Mogi case. J. Urban Technol. **15**(3), 5–24 (2008)

McMaster, R., McMaster, S.: A history of twentieth-century American academic cartography. Cartogr. Geogr. Inf. Sci. **29**(3), 305–321 (2002)

Monmonier, M.: Mapping it Out: Expository Cartography for the Humanities and Social Sciences. University of Chicago Press (1993)

Muehlenhaus, I.: Web Cartography: Map Design for Interactive and Mobile Devices. CRC Press, Boca Raton (2013)

Pickles, J.: A History of Spaces: Cartographic Reason, Mapping, and the Geo-Coded World. Psychology Press (2004)

Rashid, O., Coulton, P., Edwards, R.: Implementing location based information/advertising for existing mobile phone users in indoor/urban environments. In: International Conference on Mobile Business, 2005. ICMB 2005, pp. 377–383. IEEE (2005)

Rashid, O., Mullins, I., Coulton, P., Edwards, R.: Extending cyberspace: location based games using cellular phones. Comput. Entertain. (CIE) 4(1), 4 (2006a)

Rashid, O., Bamford, W., Coulton, P., Edwards, R., Scheible, J.: PAC-LAN: mixed-reality gaming with RFID-enabled mobile phones. Comput. Entertain. (CIE) 4(4), 4 (2006b)

Shepard, B., Smithsimon, G.: Beach Beneath the Streets: Contesting New York City's Public Spaces. SUNY Press, The (2011)

Sotamaa, O.: When the game is not enough: Motivations and practices among computer game modding culture. In: Games and Culture (2010)

Turner, A.: Introduction to Neogeography. O'Reilly Media, Inc (2006)

Tuters, M.: Forget psychogeography: the object-turn in locative media. In: Conference Media in Transition, Vol. 7 (2011)

Tuters, M., Varnelis, K.: Beyond locative media: giving shape to the internet of things. Leonardo 39(4), 357–363 (2006)

Visvalingam, M., Whyatt, J.D.: Line generalisation by repeated elimination of points. Cartogr. J. 30(1), 46–51 (1993)

Wark, M.: The Beach Beneath the Street: The Everyday Life and Glorious Times of the Situationist International. Verso Books, Brooklyn (2015)

Whitehill, B.: American Games: a historical perspective. Board Game Stud. 2(1), 116–14 (1999)

Wilson, D., Bertolotto, M., Weakliam, J.: Personalizing map content to improve task completion efficiency. Int. J. Geogr. Inf. Sci. 24(5), 741–760 (2010)

Wood, D.: Cartography is dead (thank God!). Cartogr. Perspect. 45, 4–7 (2003)

Wright, J.K.: Map makers are human: comments on the subjective in maps. Geogr. Rev. 32(4), 527–544 (1942)

Creating Shared Encounters Through Fixed and Movable Interfaces

Patrick Tobias Fischer and Eva Hornecker

Abstract Currently, our cities become more and more equipped with information and communications technology (ICT). Rarely do these systems provide a fit with everyday public life. They focus primarily on efficiency, security, safety and business. There are few system designs which support social aspects such as identification with the city and community, responsibility, everyday habits, leisure, pleasurable stay, social interaction, courtesy behaviour, play, etc.—in short, aspects of social sustainability. To outfit our future city with technology currently lacking the support of those qualities, we created several novel interaction designs to explore how to best merge ICT with the public space. This chapter presents some of our theory developed from our research-through-design approach and three case studies including suggestions for measures of success such as the number of shared encounters, average interactions per minute (ipm), or accumulated interaction time. We believe those hard facts are needed to argue for the need of playful ICT in our city that makes our public life more enjoyable.

Keywords Urban HCI · Humane city · Tangible and embodied interaction · Research through design in-the-wild · Prototyping

1 Introduction

Most of the information technology we encounter in public spaces nowadays consists of flat screens that take on the function of billboards, signage, information screens as well as mobile phone applications. The interaction model of public displays is information-push, at most, people will get their '15 s of fame' on Times Square

P.T. Fischer (✉) · E. Hornecker
Bauhaus-Universität, Weimar, Bauhausstrasse 11, 99423 Weimar, Germany
e-mail: patrick.tobias.fischer@uni-weimar.de

E. Hornecker
e-mail: eva.hornecker@uni-weimar.de

© Springer Science+Business Media Singapore 2017
A. Nijholt (ed.), *Playable Cities*, Gaming Media and Social Effects,
DOI 10.1007/978-981-10-1962-3_8

Fig. 1 *Left* and *middle* New York's Times Square and a '15 s of fame' installation. *Right* High rises transformed into a nightly spectacle in China. Photo © Martin Hornecker with kind permission

(see Fig. 1). While displays may brighten up the city and distract from otherwise ugly architecture, they do not make a city more liveable and engaging. They seldom help people to connect emotionally and socially with their city and with others. Recently, the vision of 'smart cities' has gained traction (and funding), but this predominantly focuses on sensor systems, big data analysis, with the aims of automation, efficiency, convenience and security—the citizen and urban dweller is almost non-existent in this vision (cf. Poole 2014; Greenfield 2013) aside from being the target of sensors and recipient-customer of automation.

We believe that sociality is important for social cohesion in urban spaces. Information technology design often reduces opportunities of encounters and prevents sociality to flourish even though public spaces are inherently social spaces. They are often used by groups, e.g. families and friends going to the city. Moreover, they can become a shared affair that citizens have an interest in, as something they live in and enjoy. We are also familiar with the phenomenon of the 'familiar stranger' (Milgram 1977) that we meet time over time again. Given the right occasion and a talking point, we may finally strike up a conversation. Encountering people repeatedly can slowly build up trust, enhances the chances of social interactions occurring, and can help to reduce feelings of isolation.

In our work, we explore playful approaches for the creation of *shared encounters* in the urban environment. In this, we are partly inspired by the idea of the 'playable city' as discussed for the 'Making the City Playable' conference at the Watershed in Bristol (Watershed 2014; Baggini 2014). The hope is that playful public activities foster identification with one's city, support creative appropriation, and support community and active participation.

In our own design and empirical work on playful shared encounters, we focus on interaction types from the field of tangible and embodied interaction. We explore what kinds of configurations of technology and environment effectively support playful interactions in the city space and how they provoke social interactions. In the work presented here, we focus on how to get people engaged with novel types of interactive systems in public spaces and with each other, turning these places into sociable places that support active engagement.

Usually the resulting prototype devices are content-reduced and abstract (no text or transmitted story, just, e.g. colour and light)—implementing kind of a separation of concerns, simplifying the design space (this also makes it easier to discern the attractiveness of interaction itself). We also work towards utilizing quantifiable concepts such as shared encounters in order to measure positive impact in a more objective way.

2 Background

In this article, we focus on projects that take a playful approach to fostering shared encounters in urban settings. The term 'shared encounter' builds on Goffman's (1966) observations of 'Behaviour in Public Places'. Willis et al. (2010) defined these as "[...] the interaction between two people or within a group where a sense of performative co-presence is experienced by mutual recognition of spatial or social proximity", i.e. people acknowledge each other's presence. A slightly more restricted definition by Schieck et al. (2010) limits those to "an ephemeral form of communication and interaction augmented by technology", that is, the encounter is facilitated by some technological augmented interaction.

Thackara (2005) revives Ivan Illich's notion of conviviality when he reminds us that a sustainable city needs to be "*a city of encounter and interaction*", where community is created through co-presence and the shared meaning which emerges from interacting with others in meaningful activities. The notion of shared encounters builds on the philosophical background that Thackara touches upon (Ivan Illich, Martin Buber), but in the absence of social organizations, puts its hope into everyday encounters via a 'talking point' that creates a social permission and invitation for talk. Shared encounters create a short intermezzo within our habitual routines, a dérive from the routines of how we use urban space and interact with other city dwellers. Playfulness can be one strategy for creating shared encounters, as we demonstrate in this chapter.

Few would dare to question the overall value of play nowadays, as play has been argued to be essential for learning and socialization (Brown and Vaughan 2009) and Huizinga (1955) categorized mankind as a playful creature, 'homo ludens'. But a playful approach can also help people to rediscover their surroundings by changing perspective. Precursor to attempts for a playful city can be found in the Situationist dérive and modern parkour. Place (the lived and experienced space) is changed through the interactions and social practices taking place in it. For this, Thackara (2005) refers to Malcolm McCallough, while de Souza e Silva and Hjorth (2009) trace this understanding to the philosopher Lefebvre (1991) who explained that 'place' is constructed.

There is a new interest in the value of play in the context of urban living. The notion of the 'playable city' is a counterpoint to the narrative of Smart Cities, emphasizing serendipity, hospitality and openness instead of efficiency, and offering permission to play to the public (Watershed 2014). In 2008 Droog organized an event on Urban Play in Amsterdam, showcasing playful design interventions for the public that constitute a new form of (often guerrilla, e.g. non-commissioned and permissioned) urban art (Burnham 2008). The Watershed's Playable City program, which culminated in a conference, commissioned various playful interventions and workshops that invited the entire city of Bristol to play and re-appropriate public spaces, e.g. with a giant water slide. At the 2014 Playable City conference, speaker Tine Bech emphasized that "physical play creates social bonds", and play designer Holly Gramazio described the value of play in public as "a great way to feel at home in a space, to experience it and to have a different perspective and feel you got some ownership of it". A guardian article summarized these and related arguments in the headline 'the city that plays together, stays together' (Baggini 2014).

Huizinga (1955) and most of the literature that followed him define play as separated from (serious) everyday life via the boundaries of the 'magic circle'. But this separation is increasingly challenged, in particular in the context of pervasive games and earlier movements that intentionally blur the boundaries between game/play and ordinary life (de Souza e Silva and Hjorth 2009; Montola et al. 2009). If we want to understand how urban spaces can become playful spaces, it may thus be useful to focus on 'casual play' (de Souza e Silva and Hjorth 2009), which helps people to immerse in and rediscover physical space. While pervasive games are one radical strategy for creating a different experience of the city and appropriating it, the playable city approach did not go as far, and picks up on game genres that are familiar to everybody and have less risk of tension, e.g. childhood games, group sports, performance, the new games movement.

3 Mutual Influences of Architectural Configurations, Public Life and Interaction Design

In some of our prior works, we have investigated how different physical setups of public installations (in museums) affect engagement and social interaction (Hornecker and Stifter 2006) and how novel interaction styles such as tangible, touch, and gestural interaction influence the user experience (Hornecker and Buur 2006), in particular, social experience. Given our systems are embedded in real spaces, and—in the case of media façades and urban installations—in large architectural scale spaces, it is important to think spatial, both in terms of dimensions, distances, and in terms of embedding, interactions and balance with the given environment (Fischer and Hornecker 2012). In particular, space can orchestrate human experience and social interactions, e.g. by modulating distances between people and their spatial configuration, which in turn may send out social signals (Hornecker and

Fig. 2 The movable interface SMSlingshot by VR/Urban. *Right* person shooting a message to a façade

Buur 2006; Kendon 2009; Marshall et al. 2011). Our approach to analysing urban interactions is influenced strongly by urbanism (see: Whyte 1990).

Much of the design and analysis work described in the following sections of this chapter are based on our experiences with the media intervention SMSlingshot (Fig. 2) and influenced by our Urban HCI Space Type Model (Fischer and Hornecker 2012), which we will now focus on.

Our model of space types and basic influences of architectural configuration (see: Fischer and Hornecker 2012) emerged out of experiences with exhibiting the SMSlingshot, a media-arts installation by the VR/Urban art and technology collective (that Fischer is part of). The idea of the SMSlingshot is based on the street art tactic of tagging with paint bombs. Users type messages on a keypad embedded in a slingshot and then aim and 'shoot' their message onto a façade projection or media façade. The interaction design of the SMSlingshot was already an iteration, based on experiences with a prior public intervention called spread.gun (Fischer et al. 2010) which had festival visitors type their messages at a public city terminal and then use a cannon next to the terminal to aim and shoot. Here, visitors began to queue at the screen and barely talked. The slingshot was intended to make people move around, and aimed to create a more flexible and dynamic situation in which people are not hampered to mingle, make people feel subversive, and to foster expressive gesture.[1]

The SMSlingshot was exhibited in a variety of situations and locations around the world, mostly urban art or music festivals, but also in everyday situations. The team thus experienced the influence of diverse setups and settings on how the SMSlingshot was used, how the crowd behaved, how the public space itself came to be transformed, and also saw other installations at the same locations. We then began to utilize this to analyse how people appropriate and use space, and how architecture and public life interacts with the installation and with interaction.

A basic distinction of settings is between plaza and walkway, which create different situations. On plazas (city squares and civic places) people tend to meet with others, relax and often spend time. This means that plazas are suited for narrative structures that extend over time. A walkway on the other hand is characterized by a steady flow of people. Often, it is not possible to stop for an extended time, as one would become a hindrance for others. Thus, walkway situations lend themselves more to ad hoc, short interactions.

The spatial setup and setting influences visibilities, but also how people congregate in a space, where people position themselves for activities, and how the setting is experienced. We have summarized our insights into the space type model described in Fischer and Hornecker (2012) and shortly describe four of the seven relevant space types. The *Activation Space* (*AS*) encompasses all areas from where a display as well as the activity of a 'performer' (performer display) interacting with the system can be seen, resulting in an awareness of what is going on, but where one cannot yet interact. In the *Interaction Space* (*IS*), people actively interact with the installation. In the case of the SMSlingshot, the IS moves around with the person holding the slingshot—for the spread.gun it was fixed. The *Potential Interaction Space* (*PIS*) then comprises all positions, from where people can potentially interact—with the SMSlingshot, this is everywhere the device is still being picked up by the receiver and in direct line of sight to the façade. A *Social Interaction Space* (*SIS*) emerges where people congregate and are attracted by the system, which creates an opportunity for shared encounters. These can be created among *performers*, *participants* or *observers*. With the SMSlingshot, we observed that typing a message serves as a gestation point for social encounters, as people often discuss what to type, ask others about the device, or help each other, e.g. carrying their bag so they can use both hands.

Another important aspect of the SMSlingshot is its tangible nature, which calls for embodied interaction. The movable and untethered nature of the device means it can be easily handed over, supporting fluid shifts of control and allowing for shared use by, e.g. two people. It further integrates in natural group configurations keeping core requirements for social interaction intact. The metaphor of the slingshot is easy to understand. Users find the act of shooting a message very satisfying, as a combination of the bodily act of throwing and the sound of the rubber band.

---

[1]Note: We prefer to think of the SMSlingshot and our other systems as 'interventions' and not as installation. The term 'installation' indicates a static and immobile system and we rather aim at fluid urban interventions that create a dynamic situation.

The bodily experience of shooting with an oversized device not only reminds of child's play, but also carries elements of subversive, playful rebellion. We have further found that the act of shooting is highly performative and expressive, as the posture one has to take on is very visible and stands out from the crowd. Typing the message on the other hand is local and half-private, and only direct bystanders can see the screen of the device. This combination appears to have the effect of lowering thresholds for participation, while creating some social control over the content of messages.

The concept of 'people as display', resulting from the performative interaction design of the SMSlingshot, is another lesson related to our space type model (Fischer and Hornecker 2012) that strongly influences our work discussed here. Not only the projection on a façade constitutes a display, but the user turns into a highly visible 'performer display' which attracts attention. Additional 'people displays' generated by an intervention are the 'participant display', often mimicking the gesture of shooting without actually having the interface, and the 'observer display' (e.g. observers looking in one direction), providing cues for understanding what is causing certain effects.

4 Playfully Appropriating the City

In the following, we discuss a set of experimental projects that we conducted with our students to explore urban playful interaction. Most of these were based on 'content-reduced interaction' in term of utilizing abstract content (e.g. light and colour). This allows us to focus on interaction design decisions, and to simplify the development process. Moreover, it eases evaluation and analysis, reducing the number of factors that influence audience reactions.

A crucial question to answer when creating prototypes of public installations is: How to determine success? How many people should an installation attract into interaction and how many shared encounters should be generated? To have a benchmark for comparison, we studied a well-liked permanent installation in our home town Weimar, Germany. We next describe this benchmark study, before moving on to the experimental temporary projects.

4.1 A Benchmark Study of an Interactive Fountain

On a large square in Weimar (Herderplatz, GPS: 50.981213, 11.329813), a partially pedestrian zone with high foot traffic, but also used as a throughway for inner-city traffic, an interactive fountain installation is located. The principle is simple—one jumps on a block of stone to release a water jet. Only the smallest of the three blocks visible in Fig. 3 has this function. Depending on the pressure exerted, the water squishes higher. This installation is very popular, in particular with children.

Fig. 3 Interactive fountain in Weimar and 2 children playing

Sometimes two children collaborate or an adult helps smaller children by adding their weight. Frequently, a crowd of observers gathers. We chose it for systematic analysis based on informal observations. Being a well-liked and frequented installation, it provides us with a benchmark of how a successful installation performs in attracting people and generating shared encounters.

A team of our students was instructed with the Urban HCI space type model and the role types of performers, participants, and observers, based on Sheridan's Performance Triad Model (2005). Knowing a priori about the space type model in combination with the role types helped the students to define areas for which to count people. Furthermore, it helped them to develop a notation scheme for their counting sheet. In a first step, the space types such as SIS, IS, PIS were drawn on a map in the lab. Other space types, such as Comfort Spaces, AS required a site visit to sketch exactly. The site visit and a first pilot study of two 1-h time slots refined the sketched spaces and the counting sheets. To get a baseline, total passers-by were gate-counted. For the refined study design five 1-h slots, starting from the time children are on their way to school at 6:40 a.m., to the closing time of local shops at 7 p.m. and night-time 11 p.m. The students trained another group of students in taking the role of counters with the notation scheme that categorizes individuals according to age, group membership and the role they take (performer, participant or observer) in 5-min quantified time chunks. Overall, the interactive situation was observed by two researchers per time-slot for 9 h 40 min in total, counting a total of 3586 people. More people were interacting in the afternoon when school closes and many finish work and return home or go shopping.

As expected, given the plaza is a central passageway and the fountain is located right next to one of the incoming streets to the inner-city, there were many passers-by's that did not stop. 81 % of all people just passed by. About 17 % gathered in the PIS. This number splits up in 6 % of 'performers', people actively jumping on the stone, and 11 % of 'participants' who stood right next to them, mildly engaged, mostly socially interacting. At the edge of the PIS, there was a smaller number of passively engaged observers (1 %). The low percentage of observers is a result of only counting those at the edge of the PIS as the plaza was too large to count observers for the full Display Space (DS) of the installation. Finally, 1 % (approx. 38 people) of the total people counted experienced a shared encounter with apparent strangers that began talking or playing together. From this we recognize that actual interaction numbers are not high (Fig. 4).

Yet still, the installation has an effect on the plaza, and we know that some people play with it every once in a while. This means, that it may be unrealistic to expect much higher uptake with a digital installation. In public spaces, most people simply have a goal, may be in a hurry, or be preoccupied. In fact, low interaction rates have been found to be normal in public outside spaces and are much lower for mobile phone interaction with public displays. For example, Schoeter reports an average of 0.03 interactions per minute (ipm) with a public display in Schroeter (2012) similar to Linn et al. who also report 0.03 ipm (Linn et al. 2011). The ipm for the fountain is 1.08. This ipm value is a typical average for everyday-life situations for situated interaction. Situated interventions such as SMSlingshot with 2.4 ipm and LASER Tag (Graffiti Research Lab 2012) with 2.3 ipm show higher values, but were presented mostly in the context of media art festivals. This means, that for the eyes of the observers present on a plaza, a lot of action per minute is generated from situated installations, whereas for mobile phone interaction (as in the above mentioned examples), one has to wait approximately 30 min to observe an interaction and only gets limited 'performer display' that often is not very interesting to watch.

Counter to our preconceptions, we found more adults (55 %) to be interacting than children (32 %). Another 9 % of performers were older adults, and 4 % were categorized as teenagers. We further found that 84 % of those that interact are in groups. Being in a group appears to lower the threshold for interaction. Thus, it might be a good design strategy to design for observers (so that observation is interesting and that observers can play a role) as well as for multi-user interaction, supporting group dynamics.

4.2 Kick-/Flickable Light Fragments

In the Kick-/Flickable project (Fischer et al. 2014), we explored the potential of an interface cluster of formally similar interactive objects. From the SMSlingshot, we had learned about the benefits of movable interfaces, which create an elastic PIS and SIS that can move, vanish, form and re-form. With multiple objects we further enlarge the IS and create multiple access points (whereas just one object monopolizes interaction and creates a bottleneck). Moreover, we experimented with a different interaction style—interacting with your feet. We were further inspired by related work exploring the notion of 'light bodies' (Seitinger et al. 2009) and configurable pixels (Seitinger et al. 2010).

This project developed a family of objects that each reacts slightly different to physical interaction, in particular to being kicked (Fig. 5). The project group spent considerable time building differently shaped foam prototypes and exploring different travel characteristics for foot manipulation. Some could be easily

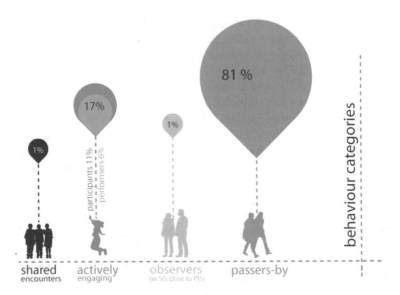

Fig. 4 Interaction count study around the fountain

appropriated as a 'football' or 'beer-belly', others provoked different interactions. Of the numerus forms, five shapes were selected, of which three were technically implemented. Moreover, each object was 'equipped' with an individual character, reacting with a different light behaviour to manipulation and proximity to other light fragments. This lends personality to the objects.

We placed the light fragments in a highly frequented pedestrian zone that also serves as a major connection path and observed for around 75 min at dusk on a warm October evening, after shops closed. From 307 people that passed by, 10 groups interacted with the light fragments, 16 people took the role of the performer while 10 were participants. This amounts to 8.5 % of people are actively engaging (5 % performers directly interacting) of all people passing-by. While this may first sound disappointing, our interactive fountain study informs us that this is not a bad result. Individuals at most made a detour to inspect the light fragments. Performers were all in groups, and frequently the rest of the group engaged as participants or distant observer, similar to what we found with the fountain. People of various ages interacted, from children over teenagers and young adults to middle-aged people, but young people (up into their 20 s) were the majority. The majority of interaction sequences lasted between a few seconds and 2 min. Most interactions were done with feet, indicating that our design was successful in terms of having clear affordances and invitations for foot interaction (Fig. 5).

Many people that went past the objects turned their heads to look at them. Observations and overheard conversations indicate that they felt compelled to sense-making, commenting on and describing the light fragments to each other. Depending on the time people spent, we can identify several levels of engagement, starting from

Fig. 5 *Left* The kick-/flickable light fragments. *Top* Two women kick one of the fragments back and forth. *Bottom* A couple walks past, inspecting the fragments visually, a child begins to explore them, and creative appropriation by a young man who picks a fragment up and puts it under his shirt

noticing and reacting (with a glance) and subsequent attempts to make sense of the light fragments, first by visual inspection and then via active manipulation. If people persisted beyond the first nudge, they would continue trying to understand the reactions, and comment on these or express their assumptions aloud. These steps of engagement so far are fairly similar to other engagement models for public installations, such as the audience funnel (Michelis and Müller 2011). But if people engaged longer with the light fragments, they began to appropriate them in creative play and to include the environment into their play, similar to the observations by Seitinger et al. (2010). For instance, one man picked the smallest light fragment up and put it under his t-shirt, imitating a colourful blinking pregnant belly, and a pair of young people began playing football, using nearby benches as goals. This integration of the environment into creative appropriation is specific for our system, as it features movable objects (Fig. 6).

Yet in some respects, we felt that the Kick-/Flickable light fragments did not yet constitute a really successful installation. While having three different objects worked to support group interaction, at several times we observed other groups waiting at a distance until the previous group left, to then quickly approach. We also observed this pattern with another installation and call it 'hidden queueing'. Many people did not appear to realize that the light fragments were interactive. The reactions were not legible enough to be distinctive, and required more time for investigation than what most passers-by were willing or able to invest. And while some people had no hesitation to manipulate the Kick-/Flickables, others seemed to hesitate, possibly being afraid to touch (and possibly break) these beautiful objects that so clearly stood out from the normal pedestrian area. Moreover, no shared encounters were generated. Furthermore, the ipm of 0.18 was relatively small compared to the fountain. But this was partially due to a couple playing with the light fragments for about 20 min. If we only consider the time span before these two arrived, then the ipm is 0.2.

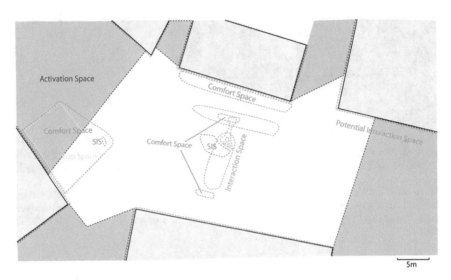

Fig. 6 Distribution of the urban HCI space types

4.3 PIPE—A Fixed Walkway Interface

With the PIPE project we explored a setup that contrasts on many aspects with the previous project. The initial idea was to utilize repeating architectural elements to increase the installation's DS and support people's ability to predict cause and effect. The custom-made display was designed to fit in a parasitic manner to rainwater pipes, which are thus augmented. Again, we decided to use coloured lights as output, and foot interaction for input. But in this case, input and output space were split up, enhancing long-distance visibility (= large AS) of our system, but sacrificing movability and with it a large PIS in return.

The student team built a light fixture that can be attached to a rain pipe, made from a chain of individually controlled multicolour LEDs. The individual segments are held in third-circle shaped elements that diffuse the light. For interaction, we had experimented with pneumatic pressure and built three tubes that people can step on.

The input mechanism thus has resemblance to the Dance Dance Revolution mats, but makes the air pressure mechanism noticeable. We wanted people to feel the tube, as this adds to the playful experience and also provides feedback of other users movements. Initial experiments with tubes resembling water hoses or bicycle wheel tubes indicated that most people would avoid stepping on these and that they could furthermore create a tripping hazard. We thus built the pressure beams depicted in Fig. 7, which create a small elevation that feels soft and a bit wobbly due to the internal air pocket. Their custom shape is created from rubber sheet material (thickness 1.5 mm), which is glued tight to a sealed wooden board. An internal sensor tracks changes of internal pressure. The IS of a single pressure beam was 1.30 m

Fig. 7 The PIPE attached to a rain drain of a university building at a busy walkway and the input pressure tubes (for stepping on) on the walkway

long, as this would allow at least 2 people to stand on them simultaneously without coming into intimate distance (based on Hall's interpersonal space model).

The interaction technique of PIPE resembles a colour mixing game. The fixture on the rainwater downpipe filled up with a section of coloured light each time a person steps on one or multiple of the pressure beams. One element blinks initially to indicate the position that will change, and also to attract attention. Stepping on a beam decides on the element's colour, each beam resulting in a different colour (the beam casing is blue-red-yellow, see Fig. 7) or, if several beams are engaged, the colours are mixed. After the internal timer for the colour selection phase is up, the element remains lit in this color and after a brief pause, the next element begins to blink. When the final element at the top of the PIPE and the entire PIPE are lit up, the lights suddenly flow up twice from the very ground (which has been dark so far, where there is a hidden spotlight) up to the top of the PIPE, clearing all created colours. The assumption behind this interaction technique was that people like to tag or leave their mark, but it turned out, the "Ring the Bell" (also known as 'Hau den

Lukas', or Strongman) metaphor was stronger. This resulted in people being very eager to reach the top light to see what happens. Thus, curiosity seemed to outplay tagging behaviour.

On 3 evenings past 8 pm, when it became dark, the system was set up and observed by our team for 1–2 h (4 h in total). The pressure mats were placed in a row in front of the output unit, which by vicinity creates association so people can guess the two kinds of objects are related. The installation was positioned at a wider walkway, which we selected for our trials as this would avoid a bottleneck for passers-by.

Overall, the multi-user setup worked very well. The pipe attracted considerable curiosity from people passing-by. The pressure beams were correctly identified by most people as belonging to the installation and it appeared to be obvious that they are for stepping onto. This is probably because they are unusual, but sturdy and big enough to not cause hesitation. From our analysis of 2.8 h video material, we counted a total of 302 people passing through the AS, 84 engaging with the system (= 0.53 ipm) and 216 passing-by. Of the passers-by, 50 % at least glanced at the PIPE and 48 % did not seem to take note of it (did not stop or slow down, and did not turn their head to glance at it) 2 % stopped, but then went on. The accumulative interaction time of all performers during the 2 h 40 min was 1 h 18 min without the additional engagement durations of the participants and observers.

During our test setup, 21 % of passers-by (64 people) became performers, which is a very high number based on our comparison data from the fountain. 13 other people (4 %) stopped to observe performers. Of performers, 15 % were categorized as elderly, demonstrating that playful interaction can reach a diverse audience.

People who passed by in a group behaved markedly different from individual passers-by. Similar to what we found earlier, 92 % of engaged people were in groups. From the 216 people just passing-by (not engaging), 30 % were individuals, and the rest were in groups. Of the individuals passing-by, only 2 % interacted with the system, whereas 25 % of people engaged if they arrived as a group. Individuals were also far more likely to pass by without glancing at the installation than people in groups. This can be explained by the tendency of groups to observe and react to each other, similar to the 'landing effect' described by Müller (2012).

In this case, we found 3 % of all people passing-by had the chance to end up in a shared encounter. The average interaction time of a performer was 1 min 13 s, which is rather long as from our experience typical interactions usually are between 5 and 30 s.

To get an impression of the variety of playful interaction (Fig. 8) nurtured by curiosity and play, we shortly describe selected observations. This also provides an impression of different performer displays generated.

The lowest three elements are on and a couple passes by. They are at least middle-aged and very properly dressed: he wears a short white shirt and long trousers; she wears a dress and white heels and carries a handbag. They have seen the PIPE and now glance down at the pressure beams. He steps on one, while she watches. Then he walks along the beams, while she watches from a few steps distance. He begins to walk a little circle, and she joins him, making small explicit steps. He walks to the side, and she begins to make small trampling steps on one of the beams, walking forward to the next beam. He walks behind her, imitating her movements. Past the third beam, she turns around, looks at the PIPE, while he is still stepping. She reaches her hands out to him and they begin to hold hands, facing each other, while stepping each on one beam, almost as if they would be dancing. They turn around each other while she looks up the PIPE again, where more and more lights have turned on. She walks to the side again and lets her partner continue to walk on the beams, watching. Then she returns to him, and walks around him in a little circle. They look at the PIPE more often now, as the line of lights comes close to the end of the PIPE. As the last light goes on, they move to the side, and she claps three times into her hands. The couple has interacted for 1.55 min with the installation (thus slightly longer than the average interaction duration). They have not filled all elements though, thus missing the finale.

The next example interaction features a group of four young people (2 men, 2 women) who quickly decipher how the PIPE works. The PIPE is almost fully lit up, as this group of four approaches, walking at a fast pace. They arrive briefly after another group has left, and thus might have seen something interesting going on. Two are carrying bottles, and they are wheeling two bikes along. One of the women walks ahead, and slows down to look at the PIPE. She stands on one of the beams, and jumps up and down, explaining "I jump", then moves to another beam and steps on it. One of the men joins the woman and they both jump. The PIPE suddenly lights up fully and the lights flow over it (the finale). The group watches: "OH!". The PIPE goes dark again, and the first element blinks. The man begins jumping on the beam again. The woman says "we need to (rest incomprehensible due to traffic noise and sound of steps)". A passing woman with a baby buggy who saw the PIPE lighting up stops to watch the group. Once a few more elements are lit up, she leaves. Now, one of the men stands with spread-out legs on two beams and jumps up and down to get the next element to light up (possibly trying to mix colours). A woman also jumps on the third beam. The other man joins and the three walk in a circle on the pressure beams, laughing. The other woman parks her bike and joins them. Their attention is focused down on the beams, but one of the women keeps looking up to the PIPE. They stop and bend down, inspecting the pressure beams and then all press down on the beams (apparently trying to maximize pressure exerted). Then they get up again, and three group members line up on two of the beams with wide-spread legs while looking up at the PIPE, stepping repeatedly left-right and shouting the rhythm. So far, they have created a pattern of a few blueish-green elements, a red, several blueish-green, another red, and more blueish elements (unfortunately only red is distinct in our video material). As the last element blinks, they pause and move to the other beam. From their behaviour, it is evident that they want to create a colour pattern and now work to create another red light (tagging behaviour). They all stand, watching up at the PIPE, and then all together jump on one of the beams, shouting 'one–two–three'. They laugh as the blinking light becomes a constant red, comment 'fantastic' and begin to leave. This group interacted for 3.05 min.

From the last group's behaviour, it is evident that they begin to understand how colours can be mixed, from initial slow experiments by two group members to concerted actions, e.g. all exerting pressure on one beam, or pressing on two or three

Fig. 8 Content-reduced installations can create a variety of 'performer displays'. *Top left* Team playing, standing in a row on one beam. *Top right* Middle-aged couple dancing. *Bottom left* Two performers running on the spot. *Bottom right* Two performers standing on two beams simultaneously to mix colours with a group of strangers observing

beams simultaneously. At the end it becomes clear, that their plan has been to create and leave behind a colour pattern. The first group (the middle-aged couple) did not appear to fully understand how colours are mixed, but still enjoyed the interaction.

In this case, a fixed interface was more successful than our previous movable and distributed system of the light fragments in terms of attracting many passers-by to interact for a considerable time. Both systems create a large PIS, allowing several people to interact. However, in contrast to the Kick-/Flickables, PIPE has an obvious and shared goal of making the light grow, which enhances motivation and encourages collaboration. It also impacts interaction durations, as with PIPE people play till the goal is reached.

Fig. 9 The Meiningen theatre-machine during setup, showing all three sections placed around the fountain (photos © Anke von der Heide)

4.4 Meiningen: Playful Interfaces in a Total Situation

Our next case study took place in the somewhat different context of a spectacle event, but also in the semi-public space of a city. Our University was approached by the city of Meiningen to prepare an event for the centennial celebration of the founder of modern theatre, Duke Georg II. Our idea included a façade mapping that enables visitors to interact with the projections. This was realized as an interdisciplinary project involving students and supervisors from MediaArchitecture and from Media-Computing [see for details Fischer et al. (2015)].

The aim was to engage the citizens of Meiningen with their city heritage, inside the historic courtyard of Georg II's castle Elisabethenburg. The castle has a curved façade, which was selected for the projection, as this provided a unique challenge for projection mapping, as well as potentially increasing the immersion effect. Different from the examples discussed so far, this installation was content-rich. In this case, the topic of the event governed content selection and we also had access to a large archive of historic images (Figs. 9 and 10).

On a technical level, the project combines the technology of façade mapping (or 'spatially augmented reality') with the interactivity of interactive media façades (Daalsgard and Halskov 2010).

Our design challenge was how to make the façade projection interactive in a way that would make it easy for people to interact, that would be inviting and enable active participation for as many people as possible, while fitting in with the entire space and with the story of the celebration. The spectacle consisted of distinct phases, starting with visitors entering the courtyard. Then, a 12 min projection show began. This was followed by children from city schools dancing in (3–4 min), carrying lighted objects that they positioned around a central fountain inside the courtyard. Once all objects were positioned, a previously dark structure lit up: the 'theatre

Fig. 10 The Meiningen theatre-machine at night with the projections, with the rope interface visible at the front (photos © Anke von der Heide)

Fig. 11 People of all ages had no hesitations playing with the projections using the rope interface (photos © Hesam Jannesar)

machine'. Now the audience could use this machine to manipulate theatre scenery backdrops from the Duke's own hands (from the museum archives).

We had aimed for a multi-user interface with the biggest possible PIS. Interaction should not be competitive and needed to be very intuitive, since the audience would need to learn how to use the interface from watching others and trying out. After some early experimentation on-site, we settled on the idea of a rope machine with which the scenery can be pulled across the wall. The machine consisted of three parts, each 3–5 m long, and controlling a different layer of imagery. It was set architecturally, being built around the fountain and making use of the existing pathways. The machine's shape was given architectural scale by adding a larger shell in the same style (Fig. 11).

Once people understood that they were allowed to use the machine (and that it was not just for children), they used it enthusiastically. People liked to explore the content of each layer (it took a while for the same images to re-appear), and generally enjoyed being able to have an effect on such a large projection. We were surprised how many mature and older citizens interacted, and how playful they were. Overall, the rope interface was very effective, and there was little hesitation to touch it. From observations and overheard comments, the low-tech design was an important factor for this, and people generally liked the 'feel' of the rope. We also observed considerable interaction across groups (i.e. shared encounters). Here it was important that each machine section had enough space along its rope for at least 2 groups (or 5–6 people) without crowding. The impulse to join in was then big enough, and even when several people had their hands on one rope, they found it easy to synchronize which way to pull. Moreover, people often began to explain to others how the machine works, and children from different groups played together. Again, we could observe the stages of engagement we have identified earlier, from initial sense-making and 'what happens if' experimentation, over systematic exploration of potential actions and their effects, to creative action in looking to achieve specific effects or, for example, holding pace with the scenery layer controlled by another section of the machine.

Overall, the low-tech and very easily observable interaction contributed to the machine's success. Here, the artistic and historic content also played an important role in keeping the audience's attention and creating delight.

In terms of numbers, the duration of the installations interactive part was in total 46.5 min and resulted in an accumulative interaction duration of 1 h 57 min. This is a 2.52 times multiplex created by the interfaces large PIS. The ipm is therefore very high with 140 performers using the rope, amounting to 3.01.

5   Conclusion

In this chapter, we presented four case studies with different situation designs that contain an interactive system and a context in a specific environment. The case studies explored fixed and movable interfaces, walkway and plaza environments and everyday-life and event contexts. To answer the question of "What kind of technology do we want to put in public places?", we propose the measure of shared encounters, a measure that can be counted and might be used to reason why a certain digital system provides value for the city and how good it is in doing so. Playful installations are one way to generate encounters. They triangulate and can create communication between citizens that are unacquainted to each other. This communication happens not just between the performers when they directly engage with the system, which might force them to coordinate to reach a common goal, but also between the other introduced roles of the participant socially engaging with the performer, giving suggestions or cheering for them, and the observers that passively engage with the action around the installation. After all, it is generating a lively

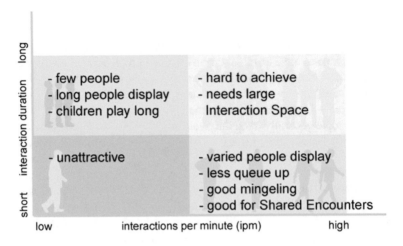

Fig. 12 Properties influencing situation designs

'theatre' to look at with entertaining value. Whyte (1990) discovered that what people like to do most in public space, is watching other people. Thus, we should always consider design for the observer. The concept 'people as displays' tries to facilitate this during the design phase.

While our studies have shown that we may be designing for a minority of people who are willing to play in public and to spend the time, this is not without effect for changing the relation to the city. Participation numbers of 5 % appear low, but many people are on their way somewhere else and may not be able to spend time. Moreover, seeing a permanent installation (like the fountain) repeatedly may create the impulse to try it out at the next occasion, and seeing others play creates the social signal that it is OK to play. Furthermore, playful designs allow a number of alternate uses. With PIPE we experienced a huge variety of 'people displays' and reactions. It is important to support these in the design as well as allowing different subjective interpretation to keep the installation sustainable (Fig. 12).

In addition to the number of shared encounters, along with the description of our case studies we have mentioned two other quantitative measures: ipm and interaction duration (also known as holding time). Both influence each other depending on the design of the system and the people present. If there are enough people present, a too long duration of interaction might decrease the value of the installation, as only few people play with it (limited throughput). They might rather queue up. For the role of the observer this does not make a big difference, as a people display is present with either few people playing long or with lots playing briefly. We have seen that ipm values between 0.1 and 1 are rather normal for situated installations and that values of 3 demand a total situation, which is event-like (the Meiningen case study). These values are good indicators for how well an installation performs in combination with multiplex factors, which indicate durations of interaction.

To increase the ipm, visibility, discoverability, and creating an interesting performer display for observers is advised. Sturdy and non-high-tech appearing input elements also seem to lower hesitations to participate. Furthermore, a large IS is also helpful to increase total interaction times.

Astonishing to us was especially the high numbers of adults, across all ages who participate and give the interactive system a try. This is against the stereotype and shows the fallacy of the assumption that these types of installations are only valuable for children. Our case studies show that older people are very well willing to engage in playful interactions.

The impulse to play is strongest in groups. Psychologically, being in a group provides some feeling of strength, and when the group 'authorizes' playful behaviour this provides permission to play. Also, in a group there are participants present so the performer has an immediate audience familiar with at the same time. For future research in public play, it might be of interest how to encourage individuals to play by reducing hesitations. Further research also needs to go into the question of "How to encourage inter-group interaction in order to generate more shared encounters?" Currently, we do not have any design factor identified which tells us why sometimes inter-group interaction happens (e.g. the fountain case and PIPE) and sometimes it does not (Kick-/Flickables). This could be because of a common goal that is easy to describe, in contrast to an explorative activity where one does not actually know what one is doing, but figuring it out on the go. With the Kick-/Flickables, the resulting effect was very much hidden and personal interpretation was strongly demanded. While a verbal externalization of one's own interpretation is easy with acquaintances, strangers might not understand each other's interpretation or be hesitant to share them.

With all our designs we aimed for multi-user support by increasing the PIS as much as possible. We also advice to support group dynamics and their fluid reformation. A physical artefact that can be handed around can increase performativity and also indicate who is playing. This is especially important for a design which aims to include the observer's passive engagement. It does provide pleasure for them, as well as explain the usage of the system. In the case of SMSlingshot, it also provided social control as inappropriate messages can be traced to the person posting it. In that way, social norms stayed intact.

With the proposed measures we do not want to undermine the importance of descriptive case studies. In fact, we believe rich descriptions may be the only way to understand why some playful situation designs for public spaces are more successful than others. However, some stakeholders can only be convinced by numbers. It is in that regard why we have to gain an understanding of what measures might be suitable for that task. Our cities need more playful digital situation designs. We have chosen to explore possible futures and encourage others to test their designs in-the-wild as well.

References

Baggini, J.: Playable cities: the city that plays together, stays together. The Guardian, 4 Sept 2014. http://www.theguardian.com/cities/2014/sep/04/playable-cities-the-city-that-plays-together-stays-together (seen 14 Sept 2014)
Brown, S., Vaughan, C.: Play: How it Shapes the Brain, Opens the Imagination, and Invigorates the Soul. Avery, New York (2009)
Burnham, S.: Urban play: an introduction. In: Burnham, S. (ed.) Droog Event 2. Urban Play. Droog Publisher, Amsterdam, pp. 8–11 (2008)
Daalsgard, P., Halskov, K. (2010). Designing urban media façades: cases and challenges. In: Proceedings of CHI' 10, pp. 2277–2286. ACM (2010)
de Souza e Silva, A., Hjorth, L.: Playful urban spaces. A Historical Approach to Mobile Games. Simulation and Gaming, vol. 40, no. 5. pp. 602–625 (2009)
Fischer, P.T., Hornecker, E.: Urban HCI: Spatial aspects in the design of shared encounters for media façades. In: Proceedings of ACM CHI' 2012, pp. 307–316
Fischer, P.T., Zöllner, C., Hornecker, E.: VR/Urban: Spread.gun—design process and challenges in developing a shared encounter for media façades. In: Proceedings of British HCI 2010, BCS eWiC series, pp. 289–298. British Computer Society/ACM NY
Fischer, P.T., et al.: Movable, kick-/flickable light fragments eliciting ad-hoc interaction in public space. In: Proceedings of PerDis' 14, The third international symposium on pervasive displays, p. 50. ACM NY (2014)
Fischer, P.T., et al.: Castle-sized interfaces: an interactive façade mapping. In: Proceedings of Pervasive Displays 2015 (PerDis' 15), pp. 91–97, ACM NY (2015)
Goffman, E.: Behaviour in Public Places. The Free Press, New York (1966)
Graffiti Research Lab: Graffitiresearchlab.com. Last accessed: 2012
Greenfield, A.: Against the Smart City. Do Publications, New York (2013)
Hornecker, E., Buur, J.: Getting a grip on tangible interaction: a framework on physical space and social interaction. In: Proceedings of CHI 2006, pp. 437–446. ACM Press, NY (2006)
Hornecker, E., Stifter, M.: Learning from interactive museum installations about interaction design for public settings. In: Proceedings of OzCHI 2006, pp. 135–142. ACM, NY (2006)
Huizinga, J.: Homo Ludens: A Study of the Play-Element in Culture. Beacon Press, Boston (1955)
Kendon, A.: Spacing and orientation in co-present interaction. In: Proceedings of COST 202 Training School, pp. 1–15. Springer, Heidelberg, (2009)
Lefebvre, H.: The Production of Space. Wiley, New York (1991)
Linn, S., et al.: Discursive architecture: integrating buildings, displays and text. In: Proceedings of CHI 2011
Marshall, P., et al.: Using F-formations to analyse spatial patterns of interaction in physical environments. In: Proceedings of CSCW, pp. 445–454. ACM (2011)
Michelis, D., Müller, J.: The audience funnel: Observations of gesture based interaction with multiple large displays in a City Center. Int. J. Hum.–Comput. Interact. **27**(6), 562–579 (2011)
Milgram, S.: The Familiar Stranger: An Aspect of Urban Anonymity. In Milgram, S. (Ed.), The Individual in a Social World (pp. 51–53). Moston, MA: Addison Wesley, 1977
Montola, M., Stenros, J., Waern, A.: Pervasive Games. Theory and Design. Morgan Kaufmann Publishers, USA (2009)
Mueller, J., et al.: Looking glass: a field study on noticing interactivity of shop windows. In: Proceedings of CHI' 12, pp. 297–306. ACM (2012)
Poole, S.: The truth about smart cities: 'in the end, they will destroy democracy'. The Guardian, Wed. 17 Dec 2014. http://www.theguardian.com/cities/2014/dec/17/truth-smart-city-destroy-democracy-urban-thinkers-buzzphrase (seen 30 Dec 2015)
Schieck, A.F., et al.: Exploring digital encounters in the public arena. In: Willis, K.S., et al. (eds.) Shared Encounters, pp. 179–195. Springer, London (2010)
Schroeter, R.: Engaging new digital locals with interactive urban screens to collaboratively improve the city. In: Proceedings of CSCW' 12, pp. 227–236. ACM (2012)

Seitinger, S., et al.: Light bodies: exploring interactions with responsive lights. In: Proceedings of TEI, pp. 113–120. ACM (2010)

Seitinger, S., et al.: Urban pixels: painting the city with light. In: Proceedings of CHI, pp. 839–848. ACM (2009)

Sheridan, J.G., et al.: Understanding interaction in ubiquitous guerilla performances in playful arenas. In: People and Computers, pp. 3–18 (2005)

Thackara, J.: In the Bubble. Designing in a Complex World. The MIT Press, Cambridge Mass (2005)

Watershed: http://www.watershed.co.uk/playablecity/conference14/ (2014)

Whyte, W.H.: City: Rediscovering the Center. Anchor, USA (1990)

Willis, K.S., et al.: Shared encounters. In K.S. Willis, et al. (eds.) Shared Encounters, pp. 1–15. Springer, Berlin (2010)

Part III
Design for Playful Public Spaces

The City as Canvas for Change: Grassroots Organisations' Creative Playing with Bogota

Leonardo Parra-Agudelo, Jaz Hee-jeong Choi and Marcus Foth

Abstract In this paper, we look at people as makers of their cities. We examine the case of Bogota where many underserved communities face daily struggles to survive. By building on Lefebvre's notion of the right to the city, we explore how civic agency, play and creativity offer fertile grounds to work towards the creative management of conflict for building active citizenship. The paper presents the results and insights of our work with two grassroots organisations in Bogota, which include four themes that bring to light how their work empowers through play, creativity and trust, the ways in which they find common ground for playful collaboration with other city constituents, the approach to street-based strategies that they use for bringing about social change, and the ways in which they work towards envisioning their future and that of the city. Finally, we discuss how conflict and difference can be leveraged to move grassroots agendas forward, and how civic agency, play and creativity are central to defining how cities are shaped by bottom-up work.

Keywords Bogota · Social change · Grassroots organisation · Urban informatics · Urban play

1 Introduction

This chapter looks at people as makers of their cities. Specifically, it takes the focus away from the wider discussion around the potential of digital and networked technologies to empower people to connect and create changes from the bottom-up

L. Parra-Agudelo (✉) · J.H. Choi · M. Foth
Urban Informatics Research Lab, Queensland University of Technology,
D Block, 2 George St, Brisbane, QLD 4000, Australia
e-mail: leonardo.parra.agudelo@gmail.com

J.H. Choi
e-mail: h.choi@qut.edu.au

M. Foth
e-mail: m.foth@qut.edu.au

© Springer Science+Business Media Singapore 2017
A. Nijholt (ed.), *Playable Cities*, Gaming Media and Social Effects,
DOI 10.1007/978-981-10-1962-3_9

189

(Foth et al. 2015). Instead, it examines the case of Bogota, a city with aspirations for economic development through creative and technology industries but at the same time with diverse underserved communities who face daily struggles for survival, and thus do not see such aspirations as relevant or meaningful to them.

In this chapter, we build on Lefebvre's notion of right to the city (Lefebvre and Nicholson-Smith 1991; Lefebvre et al. 1996; Foth et al. 2015) to consider the shift of control over the playful creation of urban spaces from capital and the state towards urban inhabitants. We perceive the city as a place where civic agency acts as driver for how urban spaces are constantly re-shaped through intended and unintended ways of perceiving, conceiving, and living by its inhabitants (Lefebvre and Nicholson-Smith 1991). Further, we draw from Jacobs's suggestion that cities have the capability of offering something for everybody when created by all (Jacobs 1961), we argue that the possibilities of transforming all the components of the city can and should rightly be in the hands of its inhabitants. However, this people-centric approach poses a challenge especially in an environment where political and civic oppressions are an omnipresent aspect of how everyday is imagined, understood, and experienced for the majority of population.

In this respect, we refer to the notion of play, as a voluntary, transient, and novel experience between the pressure of control and the possibility of autonomy and self-governance (Choi 2010), which may serve to inform notions of civic agency (see Dahlgren 2006), thereby allowing us to consider the city as a playground. Here, the inhabitants have a right to contribute with their own deeds to re-creating their city in pursuit of providing an environment that can continue to provide them with the right and capacities to live with and allow them to take advantage of plurality, diversity and equity for the betterment of their own livelihood.

This way of empowering urban inhabitants by leveraging civic creativity directly relates to the bottom-up urbanism as put forward by Iveson (2013) and the civic appropriation of the city (Choi and Greenfield 2009). Purcell (2002) argues that Lefebvre imagines appropriation as a twofold proposition: (i) as the right to occupy already produced urban spaces, and (ii) as the right to produce urban space that meets the needs of citizens. In such an instance, people can be motivated by their own needs and desires and take action, driven by the possibility of a profound change through appropriation of the urban environment and the enactment of their rights as citizens. Taking over the city via creative means can then refer to the transformation of the urban environment through open and participatory reconfiguration of physical and non-physical urban systems (see Lefebvre and Nicholson-Smith 1991), underpinned by a bottom-up ethos (Iveson 2013). In this light, the city may be perceived, conceived and lived as a fluid space that is ready to be acted upon by its people (e.g. Ferrell and Weide 2010).

The creative empowerment and action of the citizen includes multiple dimensions, which Iveson (2013) suggests can be placed across a range of various vectors of action

- Temporary to permanent
- Periphery to centre

- Public to private
- Authored to anonymous
- Collective to individual
- Legal to illegal
- Old to new
- Unmediated to mediated.

By considering these different vectors, the influence of a small intervention could range from having an impact in how people relate to the built environment (Caldwell and Foth 2014) and its qualities (e.g. Parra-Agudelo et al. 2013) to its broader economic and political conditions (Scott and Storper 2015) or in the various administrative and social city scales (e.g. Jacobs 1961) among others.

The particular case of bottom-up creativity in Latin American cities originates in the perception of how local governments abandoned their duties and evoked a general distrust among the citizens in the state's capacity to deliver appropriate services (e.g. Hirschman 1984; Rodríguez 2008, 2011) to ensure their well-being. This allowed those living under straining conditions to focus on building up their own capacities (Parnell and Robinson 2012) and embracing the value system based on self-care and mutual aid through bottom-up cooperation (Sennett 2012). It is through this approach that the lack of resources and straining conditions in Latin American cities are leveraged and result in the creative engagement to address local issues, which can have a profound transformative impact on the broader city and its inhabitants.

In this chapter, we will discuss how agency, creativity and play are tightly coupled in urban grassroots efforts; the potential of leveraging on diversity and difference; the unique urban contexts of Bogota, and; playful reconfiguration/re-creation of the urban landscape of Bogota fostered by grassroots organisations. We will then argue that civic bottom-up projects that appropriate urban environments can transform the city into a platform for launching social changes that further drive and propel individuals, communities, and organisations to conceive, perceive, and live a new, playful possible city.

2 Agency and Creativity

Latin American, in particular Colombian, cities have been at the centre of large migrations and forced displacement (Grupo de Memoria Histórica 2011). As argued by Sassen (2013), these urban centres had to develop capabilities to accommodate groups and people with diverse social, economic, and political backgrounds and agendas. Williams et al. (2009) and De Waal (2011) suggest that the congregation of people with a wide range of outlooks on life in a city means that urbanites will encounter the norm of complexity and difference and deal with it in some capacity. This situation often results in tensions driven by incompatible agendas and power conflicts. These tensions are natural and fundamental to the democratic context (Mouffe 2000) in which Colombian cities exist.

At the centre of the tensions mentioned above, lies the notion of civic agency or the individual and collective capacity to act upon everyday matters. Dahlgren (2006) argues that civic agency is a learning process in which people can acquire civic expertise gradually through trial-and-error practices. This notion serves to understand how the three dimensions of space proposed by Lefebvre and Nicholson-Smith (1991)—the conceived, perceived, and lived—can also be accessed and modified by citizens depending on how they see it fit, e.g. the physical space can be transformed by bottom-up actions such as guerrilla gardening, the conceived space instead of being in the hands of technicians could emerge from organic development such as irregular settlements, and, the resulting lived space would respond to the images and symbols that originate in patterns shaped by community decisions and actions. Thus, civic agency is central to the transformative potential that exists in underserved areas of Latin American cities.

Colombian cities have been recognised as examples of social change and the support of creativity to promote urban development (e.g. Zamudio and Barar 2013). Much of the work that is being advanced by civic initiatives in Colombia respond to the consequences of years of violence in the country and the lack of state presence, e.g. the project Mil Colores Para Mi Pueblo (A Thousand Colours For My People) that aims at recovering of public spaces through community-driven interventions in towns that fell victim violent acts (Rodríguez 2015) or the development of civic media outlets aiming at disrupting violence through community-led radio and TV stations (Rodríguez 2011). Some of these initiatives develop and deploy projects that target the oppressive and straining living conditions ubiquitous in underserved areas. Through these grassroots initiatives advanced by urban communities, the groups empower themselves by answering their own needs and desires (Rodríguez 2008, 2011), propelling approaches to address civic concerns in places where institutional presence is missing. These initiatives can be understood within the framework of creative cities proposed by Landry (2008), where creativity is a multifaceted resourcefulness that involves the capacity to assess and find solutions for unusual problems or circumstances. The notion of creative city we are referring to is grounded in civic agency (Dahlgren 2006), empowerment and bottom-up development through creativity, which adds the strong participatory aspect to the notion that cities can take advantage of diversity if designed well (Wood and Landry 2008).

At a grassroots level, Colombian cities count with an effective response to prioritising the construction of cities that correspond to human needs rather than to imperatives of profit-making, a need underscored by Brenner et al. (2009). Further, this also underlines how play challenges existential conditions that can be reformed according to human will (Choi 2010) and can be considered as a key component of civic agency. Some of these responses take into account the possibly adversarial nature of working within a diverse city like Bogota. The exploration of how creative means could foster dialogue, while considering contestation and dissent as valuable assets (Zamudio and Barar 2013), could respond to multiple perspectives about how the city should operate. By considering and supporting creativity as a result of the enactment of civic agency, urban centres like Bogota and others could benefit from a prolific emergence of civic-centric, and civic-envisioned, solutions targeting community concerns.

3   Bottom-Up Play

Play exists in a contested space between control and freedom, or between pressure from an imposed control and the possibility of overcoming that pressure (Choi 2010); and argues that overcoming oppression means that the process must necessarily involve pleasure; otherwise it would not be voluntary and could not be defined as play. Following de Certeau (1984) assertion that geo-social strategies of oppression can be actively and tactilely resisted through everyday practices, Choi (2010) suggests that dominant ideologies can be challenged and transformed through play.

Envisioning change through play requires the definition of a route outside what is established and the capacity to observe an alternative set of rules, which is a condition of play (Bial 2004). Following this trajectory, and given the liberatory character of play, Shepard (2005) suggests that the activities related to play can find a subversive social movement character and differentiates distinct elements of play as it relates to movements aimed at organising work for social change

- Social and cultural play: cultural activism
- Creative play: spaces for considering alternative solutions
- Play as performance: setting stages for social and political work
- Serious play: play that involves large social structures.

According to Shepard (2005), these elements can be used to challenge unjust social arrangements. They provide a platform from where to act upon, through direct manipulation, established cultural rules with playful solutions that exist outside traditional boundaries. The play spaces include collective and individual actions that move across the opposing conditions of oppressor and oppressed, in favour of the latter. Thus, bottom-up playful actions empower people to contribute with their own deeds in the re-creation of the spaces where they live, and in managing how plurality, diversity and equity are leveraged for the betterment of their own livelihood.

4   Urban Fertility in Difference

Sassen (2013) argues that the city is a key space for the material practices of freedom, including its contradictions and anarchies, and points at the possible threats of adversarial perspectives that clash with each other in which the powerful can take over the powerless. Mouffe (2000) adds that the friction in democratic societies lies both in the logic of equality and the logic of liberty, as equality restricts freedom while freedom undermines equality. However, a city where differences can co-exist, or an unoppressive city, requires giving the appropriate political representation to multiple interest groups and the celebration of their cultural characteristics and distinctiveness (Young 1986). This coexistence could be extended by collaborating through difference (Iveson 2013), across various creative practices (Mouffe 2007; e.g. DiSalvo 2012; Zamudio and Barar 2013) that could potentially originate in and target the urban landscape through urban play (Shepard 2005).

The case of Bogota where difference is a key component for the development of creative initiatives has distinctive qualities. During the combined administrations of Mayors Antanas Mockus and Enrique Peñalosa in the late 90s, the central administration was interested in providing avenues for the enactment of the right to the city. This right, however, only existed if it was in line with what officials considered adequate behaviour, which in turn was aligned with a global neoliberal agenda that favoured investment-worthy cities. As a result, some citizens were deemed less worthy than others, reinforcing the already existent socio-spatial fragmentation (Berney 2013).

A turning point in how Bogota approached creative difference happened after Diego Felipe Becerra, a young graffiti writer, was shot dead by police while tagging his Felix-the-cat signature and a cover-up to portray him as an armed robber was uncovered. The public outrage prompted a change in municipal policy, and the then Mayor Gustavo Petro issued a decree to promote graffiti and street art as a valuable cultural expression (Brighenti 2016).

The aftermath of these events exemplify only one case in which playful urban creative means are endorsed by the local administration. However, the recently reinstated Peñalosa government is going back to its 1990s practices of enforcing neoliberal agendas in the city (The fight for recovering public space in Bogota begins 2016) while ignoring how to leverage them for common good (e.g. Méndez Lozano et al. 2015) and proscribing the material conditions for producing and supporting diversity in the first place (Galvis 2013).

To permit and foster difference, Mouffe (2000) argues that healthy political conflicts between clashing perspectives can substitute hostile relationships between adversaries. Following Mouffe's approach Wood and Landry (2008) argue that the creative management of conflict presents an opportunity to build active citizenship, where the proposal of alternative lifestyles and the occupation of the city in every day life can foster multiple uses (Iveson 2013) of the urban fabric (e.g. Rodríguez 2015).

5   Grassroots Organisations in Bogota

Latin America is fundamentally an urban region, where almost 80 % of the population lives in urban centres (UN-Habitat 2012). In Colombia, the capital of the country has observed a significant population growth from nearly 5 million people in the early 1990s to more than 10 million by 2020 (UN-Habitat 2013). As a result of local and international investment, the city is currently classified as a business and financial service powerhouse and is listed among the top 30 emerging world cities due to its interaction and integration in the global economy (Clark et al. 2015). However, the distribution of resources is uneven in Bogota. Inequality, poverty, education coverage and limited labour markets remain as factors that have negative effects (Poveda 2011) in the city.

As a result of the suppression of civic rights by multiple agents during a long-running and undeclared internal armed conflict, Colombia has a history of grassroots organisations that deal with long and short-term consequences of it (Rodríguez 2011). The development of local approaches for dealing with social struggles in underserved urban areas of cities like Bogota (e.g. Rojas 2011) through creative initiatives (e.g. Rodríguez 2008) target inequality, racial discrimination, resource deficiencies, food justice or the lack of state presence among others. The importance of these organisations resides in the networked support they provide to people living in underserved communities or under straining conditions. In many cases, these organisations act as meeting places where citizens bring their concerns or ideas to see them addressed, supported or developed. Citizens also find social shelter in groups of like-minded people that can also offer advice informed by experience. In addition, most organisations collaborate with other groups of individuals and organisations that work around similar issues, forming a collective of networks on which the whole community can rely upon, compounding the network capacity (Chandler and Kennedy 2015) in the disruption of oppressive social structures that restrict social progress (Rodríguez 2008, 2011).

Building on the significance of the work currently being advanced by grassroots organisations in Bogota, this study examines how two organisations operationalise civic agency through playful creative means in Bogota. Our study seeks to identify how grassroots organisations make use of the city as a platform for building resilient communities and advancing ways of living based on creativity and play. In addition, we want to contribute to understanding the complex lived realities of citizens in the global South while interrogating real-life experiments of city-building (Parnell and Robinson 2012) undertaken by grassroots organisations in a major urban centre in Colombia.

5.1 Ayara

Fundación Artística y Social La Familia Ayara (Ayara) is a non-profit youth organisation founded in 1996 that addresses violence, poverty, racial, social and economic discrimination through different Hip Hop manifestations such as Graffiti, Break Dance, Rap, and Dj'ing among others (Vidal Arizabaleta 2011). Ayara has developed its own interdisciplinary High Impact Methodology for undertaking their social efforts (Young in Prison 2014; Fundación Artística y Social La Familia Ayara 2009; Vidal Arizabaleta 2011) by concentrating on local forms of Hip Hop while linking local hardships and realities to global trends and geopolitical shifts (Dennis et al. 2012). Ayara's methodology uses creative means to teach concepts such as self-esteem, conflict resolution, social skills, and purposefulness. The methodology is based on thematic and artistic workshops that integrate social and psychological components

(Fundación Artística y Social La Familia Ayara 2009; Young in Prison 2014). The organisation is interested in providing a sense of individual and collective accomplishment, supporting artistic and social growth, strengthening collective processes and individual self-esteem and the integration of local youth into their communities through the creation of tangible creative outcomes in their workshops (Fundación Artística y Social La Familia Ayara 2009; Young in Prison 2014).

5.2 MAL

Muévete América Latina (MAL) is a grassroots organisation that emerged organically around 2009. The group is driven by an interest in public space interventions. From the onset, MAL explored various forms of urban art forms such as Street Art, Graffiti and Muralism to address and highlight the struggles of underserved communities in Bogota. The group maintains a horizontal organisational structure, allowing the individual members and the group to have a significant interaction with people and organisations working on similar topics. MAL's efforts concentrate on exposing and minimising issues that include social discrimination and inequality, poverty, food justice, indigenous affairs, reclaiming public space as discussion arena, social and environmental sustainability and forced recruitment. MAL collaborates with underserved communities by running workshops and community-based collective public interventions but also works independently (Muévete América Latina 2012). The organisation has no formal documents outlining an explicit methodology for integrating community voices to their projects, although the group documents their actions on their blog (https://emeaele.wordpress.com/), and social media outlets in Twitter and Facebook. Molina Nicholls (2013) provides a personal account of how the organisation emerged and documented the early stages of the group.

Both MAL and Ayara focus their efforts in making use of art practices and the urban environment to address similar civic issues. Both organisations operate with a hands-on approach, support individual and community development, strive to achieve visible and tangible outcomes through their work in a similar fashion to other Colombian grassroots initiatives that attempt to bring about change (e.g. Rodriguez 2012) through continuous improvement (Chandler and Kennedy 2015). Moreover, both groups consider social inclusion as prerequisite for the inclusion of civic agendas and the protection of rights of disenfranchised communities and attempt to bring balance to unequal living conditions through the development of knowledge (Silver et al. 2010) that is locally relevant.

6 Methodology

To do this, our study was conducted using a combination of qualitative research methods including in situ participant observations in the form of field notes (Mulhall 2003) and focus groups (Liamputtong 2011; Stewart and Shamdasani 2007). This approach helped us immerse in the organisations' work and to build rapport and trust with the participants. We took field notes during various visits to Ayara's cultural centre that included three one-hour long visits to their headquarters where the organisation's perspectives and expectations in relation to this research project were discussed, and an hour-long social gathering to which we were invited. In addition, field notes were taken after three visits to MAL's main operation space and a work-related pushbike road trip. Two visits lasted 2 hours, and another one lasted nearly 5 hours. The road trip lasted for about 10 hours, included a visit to a family house in the town of Suesca where MAL retreats to work, two stops where they collaborated on a large mural, a lunch session that included the discussion of future activities and work, and sightseeing in the town of Guasca.

We took field notes to gain an early understanding of how Ayara and MAL situate their interests and the issues and struggles they address in Bogota and its surrounding areas. The field notes provided early insights about how both groups operate make use of their workspaces and execute their interventions in the city. The different activities undertaken during the events preceding the focus groups aimed at building trust and an increased degree of comfort in the interactions between the organisations and some of the prospective participants with the research team.

A review of the web and social media presence of both groups provided a general picture of how they communicate with their members, the communities they work with and others. It also provided information that did not emerge from the focus group data.

Table 1 Participants and their role in each organisation

Participant	Affiliation	Role	Focus group
PM1	MAL	Founding member	F1
PM2	MAL	Founding member	F1
PM3	MAL	Volunteer	F1
PM4	MAL	Volunteer	F1
PM5	MAL	Founding member	F1
PA1	Ayara	Founding member	F2
PA2	Ayara	Founding member	F2
PA3	Ayara	Volunteer	F2
PA4	Ayara	Volunteer	F2
PA5	Ayara	Volunteer	F2
PA6	Ayara	Volunteer	F2
PA7	Ayara	Volunteer	F2
PA8	Ayara	Volunteer	F2
PA9	Ayara	Volunteer	F2
PA10	Ayara	Volunteer	F2
PA11	Ayara	Visitor	F2
PA12	Ayara	Visitor	F2

Fig. 1 Flyer distributed by Ayara through social networks titled Workshop: Interactive Technologies for Public Spaces

We conducted two focus groups, one with MAL (F1) and one with Ayara (F2). Each focus group lasted about 90 min each. F1 participants included three people from the core group that coordinates MAL's main activities, and two participants that bring MAL's projects to Soacha, a town in the outskirts of Bogota. F1 took place in MAL's operation centre. F2 participants included twelve participants in total. The group was composed of members of the organisation and newcomers that answered a call from Ayara to participate in the focus group. F2 took place in Ayara's cultural centre. Details of the participants and their roles in the organisation can be found in Table 1.

MAL's focus group participants were recruited first through email, and then they distributed the call through their own personal channels. Participants involved in Ayara's focus group were recruited first through email, and the call was then re-distributed by them through their own public Facebook page with a digital flyer that they designed (Fig. 1) and word of mouth. The initial work with MAL helped to inform and re-shape the following session with Ayara. This study is the first part of a two-part study that we ran with Ayara and MAL that aimed at (i) understanding how both organisations operate, and (ii) running a design workshop about digital technologies and interaction. The second part is presented elsewhere, however, it is worth noting that the organisations decided to shift the approach during the workshop, because the participants deemed that envisaging interactive projects with technologies only accessible to 'gomelos', namely wealthy people, neighbourhoods and nations, did not make sense in underserved areas of Bogota (Zamudio and Barar 2013). Such sentiment around technologies is visible in how ubiquitous technologies in other cities, such as smart phones, have very little presence in Colombia and bottom-up solutions such as selling mobile minutes are common cities throughout the country (Méndez Lozano et al. 2015). Current smart city initiatives have been pushed by the government (Herrera-Quintero et al. 2015; Ministerio de Tecnología de la Información y las Comunicaciones 2014), however, in some cases they fail to integrate these initiatives with comprehensive covering strategies in detriment of inclusive, community-oriented and social agendas (Useche et al. 2013).

We used a thematic analysis approach (Boyatzis 1998) to derive the themes, primarily from the focus group data. The field notes were then analysed in relationship to the themes and provided additional information that we have used to further develop the themes.

7 Results

The following sections describe the themes that emerged from the focus group. The key themes relate to how Ayara and MAL are structured, work and take advantage of the qualities of the urban landscape in Bogota. The field notes taken prior to the focus groups support the four key themes that we present below.

- Empowerment through Play, Creativity and Trust
- Finding Common Ground for Collaboration
- The Street as Playground for Change
- Envisioning the Future in and of the City, Together.

7.1 Empowerment Through Play, Creativity and Trust

Ayara and MAL are organisations that gear their work towards empowering its members, volunteers and other people by supporting their development and building on individual and collective interests and projects in a thematic manner. Thus, this results in highlighting the possibilities of civic agency as enacted through serious, social, cultural and creative play:

> so people can get educated in Hip Hop...in a country like Colombia it's a very good tool...
> (in Colombia) is easy to come closer to other things, not even drugs like in some other places,
> things here are tougher. You could go from guerrillas to...it's very easy, Ayara has empowered many young people to find different and alternative answers (PA3).

This thematic approach aims at developing technical and self-funding skills that are tailored to areas in the city where the straining conditions offer limited development opportunities:

(in) our workshops we try to show alternative paths and different ways to earn a living that might be different to what one could think. Mostly in certain areas of the city where options are limited, we've focused a lot on screen printing and self-funding practices (PM2).

The organisations and their members recognise changes in how they work and the scaling-up opportunities presented as a result of working with community groups in different places around the city:

I think that at one point, I don't know if we made a conscious decision…we were just a crew of vandals and only did (graffiti) tags without much ambition, but we all noticed the (organisation's) growth when we realized we could do big things…(and) Soacha was very important, it was something I've had always wanted to do: to share our knowledge (PM1).

By placing an emphasis in developing individual and community skillsets, Ayara and MAL target everyday urban struggles with playful and creativity-based problem solving-strategies, which respond to the specific conditions of each person, community, location or situation. A key strategy applied by both organisations for the development of particular skillsets or approaching groups interested in activities such as screen-printing, music making or graffiti writing, is distributed leadership. Ayara and MAL rely upon their members and volunteers and place confidence in their capacity for leading a project:

…we always try to have someone leading each (project)…(we start) from our own interests or things that each one has seen or are almost ready to go, (for instance) the project (we did) in the swamps was something that (a member) had tried (with a friend) in Cusco (Peru). We split up, and each one brings projects along…then we sit down and discuss them (as a group) but we always try to have somcone leading (PM1)

This distributed leadership and trust not only empower members of both groups to keep bringing ideas and proposals to the organisations, but also provide an extended network on which individuals and communities can build on top of it by leveraging on the existing, and developing, available skillsets. Further, the internal operational foundation of both organisations rests on the shoulders of the collective. Thus, both groups employ similar mechanisms to maintain and expand their knowledge and expertise with the intention to develop further action capacities, expand their work networks and therefore enable the organisations to scale-up and reach and engage other groups, institutions or people.

7.2 Finding Common Ground for Collaboration

As organisations that mainly work with community groups directly, Ayara and MAL acknowledge the potential of bringing a wide range of stakeholders to the field. Both groups recognise a need for scaling-up their work and are interested in positioning themselves as communication channels between people, other organisations and institutions:

...we could be the link between the government, the educational and productive sector (and the) young people from the neighborhood: young people that have the intellectual capacity to develop these things (projects for the city)...these young people see all these (opportunities) far from reach, far from their lives. We could be the catalysts, the communicating vessel between all parties (PA1).

This collaborative mindset allows the organisations to scale-up their reach within the city at a grassroots level:

...for instance (a MAL's member) started a press agency that brought us to different (civic) processes and areas in the city, (his work) expanded our horizons...Soacha increased our scope (PM2).

In contrast, gaining traction with non-grassroots organisations even when they are aligned with the groups' interests, require more determined attempts:

...for example, the (project we did with the) Memory and Reconciliation Centre happened because we chased them and chased them... (PM2).

When asked about how to bring different themes to their work as organisations, the responses highlighted the need to find common ground between different parties, which included some that are perceived as adversarial such as the police or the church. Both groups showed an interest in establishing constructive relationships between the communities, the organisations and other institutions:

...the first thing is that these simple things make you happy. How can we bring them to a neighborhood where opportunities are limited, how do we integrate the police, and the priest, and how do we integrate the community and gain support from them?...How can we bring (our work) to underserved neighborhoods and train people in something (useful)? (PA6).

Despite the perceived or real difficulties of working with certain city constituents, both organisations understand that their work does not exist in isolation and that there are many other instances that they can have access to. Such understanding informs how they approach other groups or institutions when there are conflicting agendas, when the agendas are aligned but their enactment differs, or when the agendas are aligned and there is agreement on how to execute them on the ground. Further, the deep grasp the organisations have of how their work exists in the city, helps them identify questions, opportunities and city constituents for collaboration.

7.3 The Street as Playground for Change

A key component of how Ayara and MAL manage their everyday operation lies in how the organisations read, understand and act upon (see Fig. 2) the social and physical aspects of the city. The notion of public spaces as shared places where individuals and collectives exist both in conflict and agreement is perceived as validation to the existence of both groups:

Fig. 2 Map showing MAL's interventions in different areas and streets of Bogota

Public space is an open museum, I mean, everything…the street belongs to everyone, it's an appropriation of space, our space. Some people might not see things like one sees them… we all have the right to contribute with something. You are (in the street), you exist. It's like saying: 'hey, we're here', and other people see you and say: 'if they're there it's because there's something going on, because something is happening'… (PA2).

Streets are also the place where power relationships and conflicts are highlighted and discussed. From an adversarial stance in many cases, the distance between people, exerted by top-down city constituents, is bridged through direct appropriation of urban features:

All distances are shortened (in the street), it (happens) there, with the people, where else? (MP1) …the fact of doing it (painting on the street) is political in many ways, it's going against what's established, against what you're supposed to do…I started doing urban interventions because I had many political issues and was very upset…for example public spaces need to be public, that's fundamental, people should be able to use them and feel like they belong (MP5).

This direct appropriation approach that takes place in a street-by-street and neighbourhood-by-neighbourhood fashion means that the organisations take a close look at the local social dynamics and physical qualities of the area where interventions take place. Through dialogue, collaboration and development of community and place-based interventions the organisations build collaborative networks:

...it's been great to get to know Bogota while painting, I think painting (walls in the streets) protects us a lot, if I hadn't gone painting to many places that we have gone to, why would have I gone there for?...it's very tough if you go to a neighborhood where you're not from, there are (social) dynamics that we don't know and you go there to impose things. It was great when we started to establish a dialogue with the communities, where we found common ground where we as guests could paint things that the neighborhood wanted to see, ideas they wanted to develop...thinking more about the actual places (where we paint)...like what happened in Ciudad Bolívar...it's a great experience to go and learn (about each place), we go there, we meet the communities. Each time we paint is an opportunity to get together, talk and catch up (MP5).

Although the work that Ayara and MAL deploy in Bogota is well intentioned and makes up a relevant part in the life of their members, their work is not always well received by everyone in the city due to the local negative perception of inhabiting the streets:

(graffiti) helps you survive, (but) when you're painting, even a small letter, something that you do with passion, some people call you a vandal, or they yell you're so and so, and they treat you badly...one day a man came out (of his house) and started yelling, I started running, he called me a thieve and a group of guys caught me and hit me, I ended up with a swollen face (MP3).

In contrast, the results obtained by the organisations in helping people to get out of rough street life or straining living conditions provoke positive reactions to which the organisations respond diligently:

Sometimes you're painting and people come up to you and they offer you a (paint) job, and of course, you go ahead and do it (MP3).

As a result of their attempts at transforming the perception of the street from a place for mischief into a positive creative platform, Ayara and MAL are gaining recognition in some communities:

...many people started in the (mischievous) streets, and then came here (to Ayara), when (other) kids go and talk to them, then they come to Ayara because they know they can trust us...the (other) good thing is that parents are starting to acknowledge Hip Hop as something good, something that their kids find valuable (PA2).

At the same time, both groups try to show how the streets offer prolific grounds for social change despite ill perceptions:

I think Ayara is a multiplier of knowledge: I teach you this, and you take it somewhere else... Ayara began with (Hip Hop) and many organisations approached us and learned from what we were teaching here...that's why there are multiple Hip Hop schools in Bogota...the objective is to educate, that people begin to acknowledge that not everything is drugs on the streets (PA2).

Through direct interventions on the street and by embracing the multiple conflicts that can emerge from appropriating urban features, the organisations establish a dialogue with various city constituents, from passersby, property owners to parents and other organisations. This dialogue can have both positive and negative impact, but more importantly, the result the organisations are after is to establish themselves as civic drivers that promote discussions about what civic agency means when applied through creative play. By shifting the meaning of the street from mischievous and sterile to a canvas for creative expression, the organisations transform perceived threats and difficulties into fertile grounds for addressing civic struggles through creative means. The street is the main instrument that Ayara and MAL make use of to plan, build and deliver their work in Bogota.

7.4 Envisioning the Future in and of the City, Together

As groups that explore creative expression through different means, Ayara and MAL provide a platform for the development of individual and collective interests through thematic approaches, such as Hip Hop. These approaches justify the existence of the organisations as places for the exploration of creative life paths:

> ...Hip Hop is a (creative) youth form of expression, in some cases youth that have little economic resources or that have no formal education see Hip Hop as a legitimate life choice...thanks to that approach is that the organisation came into being... (PA2).

Further, the organisations operationalise these thematic approaches and turn the limitations imposed by living in straining economic conditions into the right set of circumstances for developing social enterprises:

> ...I think it's an opportunity to go head to head with everything that's wrong in society right now, if you do (projects) in a way that spark creativity and develop various aspects...if we make them motivating, thrilling (and) challenging that'll add to their economic viability... as I said, it's already happening in Hip Hop, break dance and graffiti, that's why there are so many young people that are coming to this artistic expression. It's a tool that prevents them to take certain (destructive) life paths (PA1).

By engendering an interest in pursuing a creative career and social enterprise, the organisations' operation also benefit from their members' own development:

> ...(with MAL) we're trying to make things happen ourselves, we have chosen to live off of what we do and pull through...(we want to) be able to live off of what we enjoy (PM1).

The mutual development feedback loop created between the organisations and its members reinforces a relationship that permeates emotional, economic and social aspects of its members' life. As such, Ayara and MAL provide a support network on which to rely upon for advancing shared and individual capacities:

> ...(Ayara) is like a family, you know, Hip Hop, whether it's graffiti, rap, break dance, there's always support from everyone, it's (like) a scaffold for many, so we can reach things that we thought we couldn't. Yes, that's it, for me it's (like) a family, a backing, a good path (PA5).

The thematic approach serves various purposes: attracts like-minded people to the organisations, provides scaffolding for the exploration of alternative life paths, and, is used to transform straining living conditions into opportunities for developing social enterprise. In turn, by turning serious, social, cultural and creative play into a living option through social enterprise and directing it towards addressing straining living conditions in a networked manner, both organisations establish a web of relationships that compound the network's collective and individual creative and social enterprise capacity. In this fashion, creative projects become social ventures that further drive and propel the organisations, its members and the communities they work with forward. Thus, enacting civic agency to its fullest and addressing inequality by enabling the creation of viable life paths.

8 Discussion

Large cities that lie at the intersection of vast migrations and expulsions often develop the capacity to accommodate a great diversity of people, which when succeed such cities enable a kind of peaceful coexistence for long stretches of time (Sassen 2013). Bogota lies at one of those intersections, however, the city does not always provide for the enactment of civic rights and citizens living in underserved areas have to deal with inequality on a daily basis. Despite straining living conditions, the local population finds ways to shape the urban fabric through bottom-up and community-based organisations.

Through a thematic approach towards the development of individual/collective skillsets, Ayara and MAL encourage their members to be aware of the social, civic and physical features of Bogota, while building trust between its members and other communities. In turn, this shapes a collective urban reading lens that serves to: (i) identify the variable nuances of different civic issues depending on where they emerge, and; (ii) defines and frames the form and scale of the interventions that the organisations execute on the streets one by one. By approaching the streets one at a time, both organisations gain a deep understanding of the dynamics of local conditions and their relationships with the city at large (e.g. Jacobs 1961). The stress placed on developing capabilities to engage civic issues through serious, social, cultural and creative play in ways that respond to the natural changes of the city, results in a dynamic understanding of how the urban fabric operates and an equally fluid and flexible response (e.g. Douglas 2014; Ferrell and Weide 2010; Klanten and Hübner 2010) from the organisations and its members.

The organisations clearly identify city constituents with similar or adversarial agendas. However, both groups comprehend the inherent contradictions in drawing strict lines, i.e. their relationship with institutions like the police or the church can be conflictive on the ground, but inevitably institutional agendas converge with their own, e.g. preventing violence or addressing inequality.

Similarly, both groups enter in conflict with people on the streets as a result of their street-level interventions. The adversarial nature of some of these relationships, whether with institutions or other citizens, stimulates Ayara and MAL to find and identify opportunities for playful collaboration. The organisations expand their reach by building bridges with other groups, broadening their networks, assembling a sort of repository of networks that contributes to compound the organisations' collective capacity (Chandler and Kennedy 2015) inwards and outwards. In turn, both groups are able to provide a platform for their members to leverage that collective capacity and develop their own social ventures that target social justice, promote even distribution of resources, give voice to everyone and hold everybody accountable (Silver et al. 2010).

By understanding the city as a space prone to being intervened and the empowerment of their members, the work advanced by our participant organisations moves across Iveson (2013) axes as a result of enacting civic agency through play and creativity, and touch upon the various elements of play described by Shepard (2005) while turning a possibility of overcoming an oppressing condition (Choi 2010) into effective change through the development and support of alternative life solutions and paths. This movement responds to their understanding of (i) how the streets can take temporary or permanent interventions; (ii) how their actions link the periphery to the centre—as city constituents or spaces—(iii) how their range of projects cover public and private social and physical layers of the city; (iv) how dissent might exist in a legal haze while being relevant and legitimate; (v) how bringing about social change in the city also has to do with understanding the relationship between the old and the new, and; (vi) their awareness that the relevance of certain forms of mediation might—or not—make sense in particular contexts. This provides the organisations with a framework through which they can turn the city into a sort of canvas that can, and should, receive the results of exerting civic agency through creativity and play.

As part of our initial intention of working towards the exploration of interactive technologies, we believe that the shift of focus proposed by the organisations answers to a critical stance about the imposition of technologies, services or the deployment of digital infrastructure that does not respond to the social realities of Colombian cities (Parra-Agudelo 2015; Useche et al. 2013). However, we argue that much can be learned from the approaches undertaken by grassroots organisations where civic agency and creativity play a central role in defining how cities are shaped through bottom-up work.

9 Conclusions

In this chapter, we have discussed the main operative strategies that we identified by running focus groups with two grassroots organisations based in Bogota, Colombia. Our study provided details about the internal and external mechanisms that the groups have developed to face some of the social challenges present in the city.

There are four themes that the data analysis revealed:

- Both organisations aim at empowering their members and the organisations as a whole through serious, social, cultural and creative play. Developing individual and collective skillets that compound the organisations' capabilities and trusting each other in leading collective projects do this.

- Both organisations understand that their work does not exist in isolation. This helps them to formulate questions and to identify playful collaboration opportunities with other city constituents.
- Streets are the main instruments that Ayara and Mal make use of for addressing civic issues in a creative play fashion. Streets are at both the beginning and receiving end of the organisations' work in Bogota.
- Ayara and MAL enact civic agency to its fullest by enabling their members and the organisations to address inequality by enabling the development of social enterprises and the creation of alternative life paths.

Our findings indicate that grassroots organisations that (i) build their supporting social backbone and inner capacity through individual and collective skill development, (ii) define the principles on which they collaborate with others, (iii) make use of the street for addressing civic concerns in a playful manner, and (iv) enable the development of alternative life options, can make effective use of civic agency through creativity and play to bring about a profound urban change (Lefebvre and Nicholson-Smith 1991; Purcell 2002) that lies at the intersection of vast migrations and expulsions, contributing to their move towards a peaceful coexistence (Sassen 2013) in difference, together.

References

Berney, R.: Public space versus tableau. The right-to-the-city paradox in neoliberal Bogotá, Colombia. In: Tony Roshan Samara, S.H., Chen, G. (eds.) Locating Right to the City in the Global South, vol. 43. Routledge Studies in Human Geography, p. 152. Routledge, New York (2013)

Bial, H.: Part IV Play. Section introduction. In: Bail, H. (ed.) The performance studies reader. Routledge, New York (2004)

Boyatzis, R.E.: Transforming qualitative information: Thematic analysis and code development. Sage, London (1998)

Brenner, N., Marcuse, P., Mayer, M.: Cities for people, not for profit. City 13(2–3), 176–184 (2009)

Brighenti, A.M.: Graffiti, street art and the divergent synthesis of place valorisation in contemporary urbanism. In: Ross, J.I. (ed.) Routledge Handbook of Graffiti and Street Art. p. 158. Routledge, New York (2016)

Caldwell, G.A., Foth, M.: DIY media architecture: open and participatory approaches to community engagement. Paper presented at the Proceedings of the 2nd Media Architecture Biennale Conference: World Cities, Aarhus, Denmark (2014).

Chandler, J., Kennedy, K.S.: A Network Approach to Capacity Building. National Council of Nonprofits, Washington, D.C. (2015)

Choi, J.H., Greenfield, A.: To connect and flow in Seoul: Ubiquitous technologies, urban infrastructure and everyday life in the contemporary Korean city. Handbook of Research on Urban Informatics: The Practice and Promise of the Real-Time City, pp. 21–36 (2009)

Choi, J.H.: Playpolis: transyouth and urban networking in Seoul. PhD (2010)

Clark, G., Moonen, T., Couturier, T.: Globalisation and competition: the new world of cities. In: Center, C.R. (ed.). JLL (2015)

Dahlgren, P.: Doing citizenship: the cultural origins of civic agency in the public sphere. Eur. J. Cult. Stud. 9(3), 267–286 (2006). doi:10.1177/1367549406066073 de Certeau, M.: The Practice of Everyday Life. University of California Press, Berkeley (1984)

De Waal, M.: The urban culture of sentient cities: from an internet of things to a public sphere of things. In: Shepard, M. (ed.) Sentient City: Ubiquitous Computing, Architecture, and the Future of Urban Space. MIT Press, Cambridge (2011)

Dennis, C.: Afro-Colombian Hip-hop: Globalization, Transcultural Music, and Ethnic Identities. Lexington Books, Lanham (2012)

DiSalvo, C.: Adversarial Design. MIT Press, Cambridge, Mass (2012)

Douglas, G.C.C.: Do-it-yourself urban design: the social practice of informal "improvement" through unauthorized alteration. City & Community **13**(1), 5–25 (2014). doi:10.1111/cico. 12029

Ferrell, J., Weide, R.D.: Spot theory. City **14**(1–2), 48–62 (2010). doi:10.1080/136048109035 25157

Foth, M., Brynskov, M., Ojala, T.: Citizen's Right to the Digital City: Urban Interfaces, Activism, and Placemaking. Springer, Berlin (2015)

Fundación Artística y Social La Familia Ayara: Portafolio Organizacional [Organisational Portfolio]. In. Fundación Artística y Social La Familia Ayara, Bogota, Colombia (2009)

Galvis, J.P.: Remaking Equality: Community governance and the politics of exclusion in Bogota's public spaces. Intl. J. Urb. Reg. Res. (2013). doi:10.1111/1468-2427.12091

Grupo de Memoria Histórica: La Huella Invisible de la Guerra. Desplazamiento Forzado en la Comuna 13. Resúmen [The Invisible Footprint of War: Forced Displacement in Comuna 13. Summary]. In: Semana, E. (ed.). Grupo de Memoria Histórica, Bogota, Colombia (2011)

Herrera-Quintero, L.F., Jalil-Naser, W.D., Banse, K., Samper-Zapater, J.J.: Smart cities approach for Colombian Context. Learning from ITS experiences and linking with government organization. In: Smart Cities Symposium Prague (SCSP), 2015 2015, pp. 1-6. IEEE

Hirschman, A.O.: Getting Ahead Collectively: Grassroots Experiences in Latin America. Elsevier, Amsterdam (1984)

Iveson, K.: Cities within the city: do-it-yourself urbanism and the right to the city. Int. J. Urban Reg. Res. **37**(3), 941–956 (2013). doi:10.1111/1468-2427.12053

Jacobs, J.: The Death and Life of Great American Cities. Vintage, New York (1961)

Klanten, R., Hübner, M.: Urban Interventions: Personal Projects in Public Spaces. Gestalten, Berlin (2010)

Landry, C.: The Creative City: A Toolkit for Urban Innovators. Comedia, New Stroud, U.K (2008)

Lefebvre, H., Nicholson-Smith, D.: The Production of Space. Blackwell, Oxford (1991)

Lefebvre, H., Kofman, E., Lebas, E.: Writings on Cities. Blackwell, Cambridge, Mass (1996)

Liamputtong, P.: Focus Group Methodology: Principles and Practices. Sage Publications, Thousand Oaks, CA (2011)

Méndez Lozano, R., Ramírez Plazas, E., Páramo Morales, D.: Aspectos culturales y socioeconómicos de los revendedores informales de minutos [Cultural and Socioeconomic aspects of Informal Resellers of (Mobile) Minutes]. Revista científica Pensamiento y Gestión **37**, 286–317 (2015)

Molina Nicholls, N.: De la pasión al argumento Bogotá territorio entre el bien y el mal [From Passion to Argument: Bogotá - Territory between Good and Evil]. In: Art Faculty. Art School. Universidad Javeriana, Bogota, Colombia (2013)

Mouffe, C.: The democratic paradox. vol. xii, 143 p. Verso, London, New York (2000)

Mouffe, C.: Artistic Activism and Agonistic Spaces. Art Res. **1**(2), 1–5 (2007)

Muévete América Latina: Muévete América Latina Emeaele [Move Yourself Latin America Emeaele]. (2012). Accessed 4 Feb 2015

Mulhall, A.: In the field: notes on observation in qualitative research. J. Adv. Nurs. **41**(3), 306–313 (2003). doi:10.1046/j.1365-2648.2003.02514.x

Ministerio de Tecnología de la Información y las Comunicaciones: Cifras Tercer Semestre de 2014. In: Ministerio de Tecnología de la Información y las Comunicaciones (ed.) Boletín Trimestral de las TIC. Ministerio de Tecnología de la Información y las Comunicaciones, Bogota, Colombia (2014)

Parnell, S., Robinson, J.: (Re)theorizing Cities from the Global South: Looking Beyond Neoliberalism. Urb. Geogr. **33**(4), 593–617 (2012). doi:10.2747/0272-3638.33.4.593

Parra-Agudelo, L.: Envisaging change: supporting grassroots efforts in Colombia with agonistic design processes. In: Paper Presented at the Interplay in Design Proceedings and Program, IASDR, Brisbane, Australia (2015)

Parra-Agudelo, L., Caldwell, G.A., Schroeter, R.: Write vs. type: tangible and situated media for situated engagement. Consilience and Innovation in Design Proceedings and Program, IASDR, **1**, 4818–4829 (2013)

Poveda, A.C.: Economic Development, Inequality and Poverty: An Analysis of Urban Violence in Colombia. Oxf. Dev. Stud. **39**(4), 453–468 (2011). doi:10.1080/13600818.2011.620085

Purcell, M.: Excavating Lefebvre: The right to the city and its urban politics of the inhabitant. GeoJournal **58**(2–3), 99–108 (2002). doi:10.1023/B:GEJO.0000010829.62237.8f

Rodríguez, C.: Lo que le Vamos Quitando a la Guerra: medios ciudadanos en contextos de conflicto armado en Colombia [What We Are Taking Away From War: Civic Media in Armed Conflict Contexts in Colombia], vol. 5. Centro de Competencia en Comunicación para América Latina, Frierich Ebert Stiftung, New Delhi (2008)

Rodríguez, C.: Citizens' Media Against Armed Conflict: Disrupting Violence in Colombia. University of Minnesota Press, Minneapolis (2011)

Rodríguez, E.: Mil Colores Para Mi Pueblo [A Thousand Colours For My People]. http://www.milc oloresparamipueblo.com/nosotros.php (2015). Accessed 14 Mar 2016

Rodriguez, M.: Colombia: From grassroots to elites—how some local peacebuilding initiatives became national in spite of themselves. In: Mitchell, C.R., Hancock, L.E. (eds.) Local Peacebuilding and National Peace: Interaction Between Grassroots and Elite Processes. Continuum, New York (2012)

Rojas, N.: The art of development: images promoting dialogue and alternatives to poverty and violence in local communities of Colombia. Cons. J. Sustain. Dev. **5** (2011). doi:http://dx.doi.o rg/10.7916/D8DR2V5P

Sassen, S.: Does the city have speech? Pub. Cult. **25**(270), 209–221 (2013)

Scott, A.J., Storper, M.: The nature of cities: the scope and limits of urban theory. Int. J. Urban Reg. Res. **39**(1), 1–15 (2015). doi:10.1111/1468-2427.12134

Sennett, R.: Together: The Rituals, Pleasures and Politics of Cooperation. Yale University Press, New Haven (2012)

Shepard, B.: Play, creativity, and the new community organizing. J. Progr. Hum. Serv. **16**(2), 47–69 (2005). doi:10.1300/J059v16n02_04

Silver, H., Scott, A., Kazepov, Y.: Participation in urban contention and deliberation. Int. J. Urban Reg. Res. **34**(3), 453–477 (2010). doi:10.1111/j.1468-2427.2010.00963.x

Stewart, D.W., Shamdasani, P.N., Rook, D.W.: Focus groups: theory and practice, vol. 20. SAGE Publications, Thousand Oaks (2007)

The fight for recovering public space in Bogota begins: Comienza la lucha por recuperar el espacio público en Bogotá [The fight for recovering public space in Bogota begins]. http://www.eltiempo.com/bogota/recuperacion-del-espacio-publico-en-bogota-en-alcaldia-de-penalosa/16492026 (2016). Accessed 25 Jan 2016

UN-Habitat: The state of Latin American and Caribbean cities 2012. In: UN-Habitat (ed.) Towards a new urban transition. p. 194. UN-Habitat (2012)

UN-Habitat: State of the world's cities 2012/2013: prosperity of cities. Routledge, New York (2013)

Useche, M.P., Silva, J.C.N., Vilafa, C.: Medellin (Colombia): a case of smart city. In: Paper Presented at the Proceedings of the 7th International Conference on Theory and Practice of Electronic Governance, Seoul, Republic of Korea (2013)

Vidal Arizabaleta, M.C.: Sostenibilidad del proyecto cultural Fundación Artística y social La Familia Ayara. Una década de estrategias de management, liderazgo y emprendimiento: Estudio de caso [Sustainability of the cultural project Social and Artistic Foundation La Familia Ayara. A decade of management, leadership and entrepreneurship: Case Study]. Universidad EAN, Bogota, Colombia (2011)

Williams, A., Robles, E., Dourish, P.: Urbane-ing the City: Examining and Refining the Assumptions Behind Urban Informatics. In: Handbook of Research on Urban Informatics: The Practice and Promise of the Real-Time City. IGI Global (2009)

Wood, P., Landry, C.: The Intercultural City: Planning for Diversity Advantage. Routledge, New York (2008)

Young, I.M.: The Ideal of Community and the Politics of Difference. Soc. Theor. Pract. **12**(1), 1–26 (1986)

Young in Prison: COPOSO methodology general manual: contributing positively to society. In: Prison, Y.I. (ed.) Young in Prison, Amsterdam, The Netherlands (2014)

Zamudio, R.M., Barar, F.: Looking for the creative city: urban development through education and cultural strategies in Medellin, Colombia. In: Paper Presented at the The Idea of Creative City: The Urban Policy Debate, Cracow, 17–18 Oct 2013

Designing ICT for Thirdplaceness

Vinicius Ferreira, Junia Anacleto and Andre Bueno

Abstract Thirdplaceness is the sense of being in a third place without architectural constraints. Third places are places that host regular, spontaneous, democratic, neutral, informal, and pleasurable anticipated gathering of individuals in which people can express themselves freely. These places contrast with the realms of home and work (first and second places), having an important role in community life in supporting civic engagement and community strength. Oldenburg defined the need for and properties of third places more than two decades ago, describing them as the heart of a community's social vitality. Bars, bakeries, parks, town squares, theaters, and churches are typical examples of potential third places. In third places, third-placeness occurs often maintaining and reinforcing in the community this sense of third place. Once society and technology have changed since Oldenburg introduced the concept of third place, we describe in this chapter how to design applications for public spaces in order to promote thirdplaceness. In addition, we present and discuss two public installations—Selfie Cafe and WishBoard—used to observe the incidence of thirdplaceness that emerged through the interaction with the interactive system. In both installations, we were able to notice the essential role that Information and Communications Technology (ICTs) can play in promoting self-expression supporting, encouraging, and fostering social interaction and thirdplaceness creating a social place.

Keywords Thirdplaceness · Pervasive computing · Urban computing · Social computing · Entertainment computing

V. Ferreira (✉) · J. Anacleto · A. Bueno
Advanced Interaction Laboratory, Department of Computing,
Federal University of Sao Carlos (LIA–UFSCar), São Carlos, SP, Brazil
e-mail: vinicius.ferreira@dc.ufscar.br

J. Anacleto
e-mail: junia@dc.ufscar.br

A. Bueno
e-mail: andre.obueno@dc.ufscar.br

© Springer Science+Business Media Singapore 2017
A. Nijholt (ed.), *Playable Cities*, Gaming Media and Social Effects,
DOI 10.1007/978-981-10-1962-3_10

1 Introduction

Interactive technologies in public contexts are becoming an increasingly ubiquitous and public affair. This growing interest raises a host of new design challenges for Human-Computer Interaction (HCI), considering a more social perspective. In this hyperconnected world, public spaces once used as gathering places for the community, such as parks and squares, are getting less appealing than virtual spaces, resulting in a lack of interconnection among locals and community fragmentation (Brenny and Hu 2013). Such gathering places outside work or family-based contexts, described as third places, provide the feeling of inclusiveness and belonging to a community. Third places host the casual, voluntary, regular, and happily anticipated gatherings of individuals. Such places are the core setting of informal public life in a community. Oldenburg (1999) argues that the neutral and leveler ground in third places support to strengthen citizenship, performing a crucial role in the development of societies and communities. Traditionally, third places are a generic designation for a great variety of public places such as pubs, cafes, coffee houses, barbershops and beauty salons. However, to constitute a third place those places must exhibit certain attributes constructed through specific social and environmental character-istics, discussed in the Sect. 2.

In contemporary society, individuals suffer from the lack of places to relief from the stressful demands of work and home life, leading to a consumption-oriented culture and the absence of informal public life. Spaces for relaxation and leisure are becoming objects of consumption and private ownership (Putnam 2000). For the greater community, the third place fosters stronger community ties through social interaction, promotes quality of life and reinforces in their regulars the sense of community.

Sense of community is a perception of similarity that makes one feels as part of a larger dependable and stable structure, an interdependence that people have with others combined with a willingness to maintain this interdependence (Sarason 1986). Sense of community can emerge combining territorial and relation dimen-sions (McMillan and Chavis 1986). The territorial dimension focuses on the territory, shared space, and proximity. On the other hand, the relational dimension includes factors of nature and quality of relationships.

Aiming at strengthening the sense of community, several studies have found that the use of Information and Communications Technologies (ICT) such as public displays often associated with personal mobile devices, wireless internet available and the presence in virtual social networks (e.g., Facebook and Twitter) when contextualized in public settings can enrich the nature of the existing spaces. In addition, those technologies can enhance the community awareness, foster civic engagement, and promote socialization through face social interaction (Farnham et al. 2009; McCarthy et al. 2009; Salvador et al. 2005). Thus, understanding how to design for social experience in public and semi-public spaces is getting ever more pertinent.

In this chapter, first we describe the concept of third places and its evolution into thirdplaceness. Second, how ICT can support communities and allow self-expression. Then, we present how we approached to design two installations exploring different ways of promoting thirdplaceness using ICTs and enabling people to express themselves, and then we discuss our findings with these installations. Finally, we describe our conclusion claiming that public sharing represents a promising model for mobile social collaboration in promoting thirdplaceness and reinforcing the sense of community.

2   Third Places

Oldenburg (1999) coined the term "third place" or "great good places" to describe public places where people gather for social interaction and enjoy each other's company. Beyond the realms of home (first place) and workplace (second place), third places promote an intimate personal ties among individuals who attend such places strengthening the sense of community. However, a harmonious balance between the domestic, productive, and sociable realms of everyday life is necessary to maintain a good quality of life.

Third place emerged from the separation of social and everyday experience in three places: first, second, and third. The home is the "first place," characterized by being a domestic and private location. The "second place" is the work, environment dedicated to achieving a productive and/or paid activity. The "third place" is an environment for inclusive socialization and conversation, where life in society can flourish (Oldenburg 1999).

The term third place represents those public places characterized by being democratic, promoting informal conversations and discussion of ideas, thus forging the community profile. These places nourish relationships and a diversity of human contact complementing the social experience of individuals in a society, in addition to work and home. However, third places are not mere shelters of home and work (Oldenburg 1999). Cafes, coffee shops, community centers, bars, beauty parlors, theaters, and squares are examples of third places, once they meet certain characteristics.

In order to promote the unique social experiences associated with these great good places and to consider a place as a traditional third place, the following characteristics are essential:

- **Neutral ground**: the place should allow occupants the ability to come and go when they want, without any obligation or undertaking, be it financial, political, legal, or any order;
- **Leveler**: there can be no importance on the social or economic status of individuals, no prerequisites or requirements for participation or acceptance in a third place;

- **Conversation is the main activity**: the relaxed and humorous conversation is the main focus of activity in third places, although do not need to be the only activity;
- **Accessibility and accommodation**: the place should be economically and physically accessible, provide the necessary to make people accommodated and feel that their needs were met;
- **The regulars**: the place has regulars who help define the characteristics of the space and facilitate the accommodation of new patrons, making them feel welcome;
- **A low profile**: no extravagance in architecture and design and offer a welcoming atmosphere for people from different social classes;
- **The mood is playful**: With no tension or hostility, the tone of the conversation is always relaxed;
- **A home away from home**: the occupants often may feel belonging to the environment, feeling comparable to the warmth they feel in their own homes.

Third places are crucial components in a community, helping to create a sense of community and place. These places often invoke a sense of civic pride, diversifying and enriching local culture and democracy. However, according to Oldenburg (1999) World War II marks the decline of informal public life, mainly, in the United States. He claims that urbanization process prompted an urban renewal taking down the importance of congenial, unified, and vital neighborhoods. Unlike the old neighborhoods, most residential areas have been designed to protect people from community rather than connect them to it. Putnam (2000) also associate this decline of third places to the new media, explaining that third places are means to keep in touch with reality, giving the people opportunity to question, protest, sound out, supplement, and form opinion locally and collectively. In contrast, efficient home-delivery media systems, such as television and newspapers, tend to be one-way communication, which can make shut-ins. On the other hand, many scholars have questioned that, once new media might be part of the solution (Schuler 1994; Wellman 1998).

2.1 From Third Places to Thirdplaceness

Directly addressing to Oldenburg's third places, studies of virtual environments such as social networks, chat rooms, and multi-user systems claim that cyberspace can rebuild aspects of community working as a third place (Kendall 2002; Schuler 1994). According to Agren (1997), as a reaction to the disappearance of third places in the physical world people have found in virtual environments their third places. These virtual environments have provided "the death of distance" and opportunities for people with common interests to communicate synchronously and asynchronously in cyberspace leveraging third place characteristics (Pasick 2004). Several online social networks have attributes that describe a third place. Taking as an example the social networks Facebook and Twitter, they are affordable virtual environments, inclusive and anyone with a registration can discuss or express comments and opinions, serving different user profiles and regular. These virtual environments are a neutral ground, warm and relaxed for some regulars are kind of like a "home away from home" (Soukup 2006; Memarovic et al. 2014).

However, those virtual environments are not the same kind of place that Oldenburg had in mind, once they do not provide the "realness" of the interaction or need for simulation (Turkle 1996). According to Farina (1998), a third place exists only attached to its neighborhoods, to local work, play and family life, to the institutions and activities that encompass daily routine. In addition, virtual environments dramatically differ from third places, once third places emphasize localized community, are social levelers and accessible (Soukup 2006). Nevertheless, both public places (e.g., third places) and more private or exclusive places are crucial in building social capital (Putnam 2000).

With the absence of third places, some companies began to announce that its stores are third places, e.g., Starbucks and Applebee's. Despite that, these companies have failed in meeting some characteristics of third places and they do not engage its customers to have informal social interactions (Khermouch and Veronsky 1995). For Oldenburg (1999), third places have a civic responsibility for the maintenance and revitalization of the community. On the other hand, locations such as Starbucks emphasize customer satisfaction rather than civic responsibility, referring to an interpretation of "third-place-ness" (Walker 2010). However, this interpretation may not meet to social needs of a third place. Considering the lack of third places and the need for promoting the sense of being in a third place, we prefer to use the definition of thirdplaceness as the event of achieving the third place's characteristics in certain place and time (Ferreira et al. 2015a).

Thirdplaceness is defined as the state where and when a specific space have the features of a third place, independent of time constraints (e.g., a happy hour). Thirdplaceness is the feeling that people have in a certain place at a certain time that makes them feel, interact and express themselves as in a third place, in a process of socialization. This sense of place and time for community socialization is built and sustained through the experiences and interactions with the space at a certain time, with others who are present there, and may be physically or virtually (Ferreira et al. 2015a).

Studies from Memarovic et al. (2014) and Calderon et al. (2012) found evidences of the thirdplaceness experience, observing a spontaneous and unplanned formation of groups in public spaces (e.g., on a sidewalk, a corridor, a gateway) to socialize and discuss everyday issues. In those groups, people came in and out forming as they please: forming, changing, and ending groups over time. Memarovic et al. (2014) associate this experience with the creation of discussion groups on Facebook, suggesting that this behavior is a clue to the similarity between the way people tend to deal with others in physical and virtual sites, blurring the distinction between physical and virtual world.

Memarovic et al. (2014) and Ferreira et al. (2014b) argue that using ICT to promote thirdplaceness experience can give communities more opportunities to strength their identity and ties. Furthermore, thirdplaceness experience can transform the location in a third place for a moment or, sometimes, permanently. However, in third places the thirdplaceness experience is very often maintaining in the community that feeling of 'great good place'.

Ferreira et al. (2015b) and Calderon et al. (2012) have observed that ICTs can provide content and contextual information to leverage discussions, interactions and collaborations among people. According to Anacleto (2014), this occurs since ICT provide information to people facilitating and encouraging conversation and social interaction. In addition, ICT provide access to the desired information to promote discussion. People can have access to different ways for people to interact with each other, both locally and virtually. Interactivity can strengthen the relationship between the third place and its visitors leading to stronger social ties.

Thus, the concept of thirdplaceness evolves the crucial role that third places have spaces in society, transforming the concept of third places in a phenomenon and not a place at a certain time. We believe that urban planners and designers should discuss and develop thirdplaceness, considering the future of urban spaces, sustainability of community life, promoting interconnection among people, remembering and celebrating the culture of the community.

Third places provide means for people to connect with their community through personal ties outside the realms of home and work (Oldenburg 1999). For the individual, third places provide a sense of inclusion and belonging to the community (Soukup 2006). These factors are crucial in supporting thirdplaceness, which is the focus of this chapter. In addition, the design of applications for public spaces should consider the user experience and other variables beyond the technology, as the space where the user will use the application (Preece et al. 2002).

Thirdplaceness emerges from the experiences and interactions with and in the place, which can transform a space in a third place-like permanently or just for a certain time. To achieve thirdplaceness, designers need to promote a context for sociability, spontaneity, community building and emotional expressiveness. In addition, thirdplaceness is not limited to architectural constrains as a third place is. For example, in a barbecue or in a house party the thirdplaceness can occur, giving the people the temporary sense of being in a third place.

Aiming at creating a favorable environment to support the feeling of being in a third place defined as thirdplaceness, we describe the design of two interactive public installations in the Sect. 4. With these installations, we intend to stimulate public self-expression using public displays in a space socially abandoned by the community to promote thirdplaceness.

3   ICT to Support Community

Cities are evolving at a rapid pace bringing new challenges for urban designers, such as air pollution, increased energy consumption, traffic congestion and effective representation of citizens' interests. Until recently, these problems have been nearly impossible to mitigate, given the complexity and dynamics of cities. However, nowadays there are technological means to help at solving such problems by capturing and interpreting data produced in urban areas. Such data may produce information to help in the planning and development of cities, one of the goals of urban computing (Zheng et al. 2011).

Urban computing is a process of acquisition, integration and analysis of a large and heterogeneous set of data generated by various sources in urban areas, such as sensors, devices, vehicles, buildings and human beings. With such data is also possible to understand the nature of urban phenomena and even predict the future of cities (Zheng et al. 2014). Therefore, urban computing comprises an interdisciplinary field of research, integrating computer science with traditional fields, such as transportation, civil engineering, economy, ecology, and sociology, in the context of urban spaces.

The vision of urban computing is on the need to address the problems of big cities. For this, this research area aims at extracting knowledge from the data collected in urban spaces by technologies to benefit people who live in those cities (Zheng et al. 2014). Urban computing is an emerging and challenging field of study intending to provide new experiences to people to improve their lives while living in the urban area through computer applications.

Although urban computing is a recent research area, since the early 1990s (Schuler 1994) there has been a considerable enthusiasm on how to use ICTs to support local community ties. However, many government strategies and policies focused on providing infrastructure and public internet access points (Gaved and Anderson 2006). Paulos et al. (2004) state that mobiles devices readily connect us to friends and known acquaintances. Nevertheless, there is a lack in exploring and playing with our subtle connections to strangers in public spaces.

Information and communications technology applications are increasingly pervading our shared urban spaces. We claim that promoting self-expression through public sharing technological applications represents a promising model for mobile social collaboration in promoting and reinforcing the sense of community.

3.1 Promoting Self-expression with ICT

Self-expression is a common practice for humans in everyday life. This kind of expression occurs in several ways and involves projecting thoughts, feelings, opinions, and identity into the world. With this practice, people reveal their internal attributes, such as preferences, beliefs and values. In Western culture, self-expression is valued as a powerful sign of freedom for individuals (Kim and Sherman 2007).

For many years, people have been using public space to express themselves, either to protest and show their indignation by a subject, or to show their creativity using for example a graffiti on a wall or writing in public toilets doors. According to Brenny and Hu (2013) people are always looking for ways to express themselves, but some forms of expression can be illegal when done in a non-organized manner. Promoting organized forms using urban interactive installations are a great way to foster self-expression and to strength the sense of community and interconnection among people in a community.

In order to help people to break the social barriers several studies have explored the possibilities of ubiquitous technologies in public spaces. Abouzied and Chen (2014) explored the use of a wristband, called CommonTies, to encourage social interaction among strangers. This technological device tries to match contextual information among people and warns users when there is someone close by with same interests. CommonTie considers as contextual information data from users, such as, where they are heading to and how close they are. By having something in common, this can facilitates and encourages people to find someone and start a conversation. Aiming at proving relevant content for people content to support socialization, Kim et al. (2010) conceived a social media platform. This platform provides recommendations based on user's location, interests and social relationship information, allowing people to know better each other.

Collaboration in public spaces is also an important issue. The ActiveCampus project (Griswold et al. 2004) uses location aware systems and public broadcasting to stimulate social interactions among students in a university campus. In this application, they observed that people are more likely to message each other when they are in close proximity to one another, even with strangers. This study suggests that relative location is a relevant factor for designers in community-oriented applications.

Taking the idea of a community gardening, Bueno et al. (2014) created a system called "Watering the Garden." This system consists of a small garden enclosed into a coffee table, a projected display, a situated large display and a sofa. The main goal of the system is to engage people to interact with strangers when they are using the system to take care of the garden. With this project, they argue that introducing physical objects on an interactive system can effectively attract passersby to interact with it.

To promote community interaction and place awareness, Memarovic et al. (2012) proposes a framework called Interacting Places Framework (IFP) focusing on public displays. This framework covers the key elements of interacting places, helping designers to develop public display applications. IFP comprises the following components:

- **Content providers**: people who are supplying or posting the content;
- **Content viewers**: people who are consuming or visualizing the content;
- **Communication channels**: they deliver the content and range from public to closed-group channels;
- **Awareness diffusion layer**: this layer promotes community interaction either explicitly (through content tailored towards a specific audience) or implicitly (by observing the output for other people).

An example of a community-oriented application is an installation called "City-Wall" (Peltonen et al. 2008). This installation has a multi-touch large public display that shows pictures from Flickr tagged with 'Helsinki'—the name of the city in Finland. In this study, they observed that people, without noticing it, started to talk with strangers when the image they were manipulating went to the other part of the screen where there was someone else.

Public authoring enriches the space by sharing local information, knowledge and experiences (Lane et al. 2005). In order to improve people's engagement with their community, interactive public displays and public art installations are mechanisms widely used. Several studies show that public displays can promote place attachment, community awareness, co-located interactions and technology-supported relationships (Farnham et al. 2009; McCarthy et al. 2009; Salvador et al. 2005). In addition, public displays can foster face-to-face human interaction and encourage collaboration among community.

Transforming in playful spaces is another strategy adopted in public spaces (Silva and Hjorth 2009). By using games, the urban space can work as a game board allowing people to interact with interactive installations or buildings. Silva et al. (2014) created an instance of the game breakout enabling people to play with the game with their body. They found that using games can allow people to collaborate with strangers without realizing. For Abouzied and Chen (2014), adopting technology to introduce new practices and change the way people use a certain space does not necessarily means a problem.

Emotional expression is a critical component of social interaction and ICT can enable this kind of expression. For this, Silva and Anacleto (2015) implemented a system called "Emotifeed" that allows staffs to give emotional feedback anonymously about the announcements from their superiors. They observed a change in the organization into a more organic and flexible structure, promoting an emotional balance in the community.

Designing tools for self-expression in public spaces has an important role for people in a community, remembering and celebrating their own culture (Brenny and Hu 2013). However, the tools need to emphasize social engagement, long-term social impact, and social capital, attending people with little or no technical knowledge. For Scheible et al. (2007), creative and playful approaches can provide people a community awareness and belonging, making people feel that they are collaboratively contributing to a common goal. Moreover, to understand the role and impact of technologies to promote thirdplaceness and socialization in a public space, an increasingly adopted approach is the in-the-wild study.

In-the-wild study involves the deployment of technologies in real-world conditions to observe the use of these technologies by people in a real context of use (Rogers 2011). Rogers et al. (2007), studies conducted in the laboratory of certain technologies, especially the ubiquitous, may fail to capture many of the complexities of the situations in which applications will pass. The in-the-wild approach allows researchers to explore and understand how people understand, use and appropriate new technologies. In addition, this approach has been widely adopted by researchers in several areas, such as, Ubiquitous Computing, HCI and Computer Supported Cooperative Work. The advantage of using this approach is that users behave more naturally, making the findings more ecologically valid. On the other hand, researchers must have a concern on the privacy of users and the exposure of participants (Marshall et al. 2011).

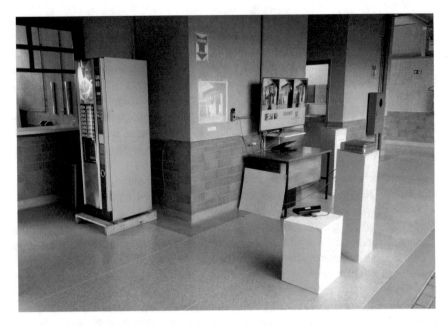

Fig. 1 Selfie Cafe installation in a common space at a public university department

In order to increase the feeling of interconnection among people of a public space, we present in the next sections two interactive installations—Selfie Cafe and Wish-Board. We deployed these installations in a (semi-)public space, using the approach of an in-the-wild study.

4   Selfie Cafe

Selfie Cafe is an interactive installation that allows users to take a selfie (a photo of themselves) and share it publicly in a large display. With this application, we intend to support, encourage and foster social interaction among students, professors, staff, and visitors of a university.

We placed the installation near of a coffee machine situated inside of a university department, as presented in Fig. 1. The focus of the application is to work as an "ice-breaker," creating a social buzz and helping strangers to interact to each other.

4.1 Installation Design

The idea of creating Selfie Cafe came from the growing movement around taking a "selfie." According to Shipley (2015), the pure and ideal form of the selfie involves taking a photo by holding the camera at arm's length, preferably using the front camera of a smartphone to compose the image on the screen or point it into a mirror. Selfies can be individual or posed with others. However, this aesthetics of the selfie have evolved to such an extent that an individual no longer have to take the photo himself or even be in it for it to be considered a selfie. Selfie is a type of self-representation of an individual with the self as the main protagonist. The ideology of selfieness is from taking a picture of the subject of the selfie in a spontaneous moment of self-production.

To support taking selfies in a playful way, we created the Selfie Cafe. This application differs from others in the way of taking photos. Instead of asking people to use their mobile devices to take the pictures, we used a toolkit called UbiDisplays (Hardy 2013). This toolkit allows creating interactive interfaces from projected displays, transforming any surface into a touchable one. We used this to let people curious to how the application works, motivating the use and adoption of the installation.

The process of interacting with Selfie Cafe and taking a picture consists of three steps. First, the user touches on the camera icon displayed on the touchable screen on the wall. After that, a 5 seconds countdown starts on the projected screen, over the image of the user captured by a webcam. With this countdown, the user can be prepared for the shot. When the countdown finishes, the system takes the selfie and the projection freezes allowing user to review the picture taken. Thenceforth, users can choose to share or delete the photo. After choosing one of these options, the system gives a feedback message to the users.

We used a large display to show publicly the selfies taken using a photo carousel. We placed this large display at the side of the projection on the wall, where people would take the photos. Every time a new photo comes into Selfie Cafe system, the large displays present it allowing people to show others.

In order to introduce some gamification element, we included a voting system for people elect the best selfie. The winners within the most voted selfie win a prize, which in our case is free ticket for the coffee machine situated at the side of the system. To allow people voting, we provided a QR code and a tiny URL. Moreover, we presented on the large display the three photos most voted to make people aware of the competition status. Therefore, even if people do not want to take a selfie they can participate of the social activity by voting or talking about the shared photo.

The system was developed using C#, PHP and JavaScript and we used the MySql database to store the photos that were converted to base64 data, to better manage the access to those pictures. We also used a Microsoft Kinect Sensor in order to capture the users' movements on the wall to make it touchable with support of the UbiDisplays toolkit.

4.2   Discussion

Aiming at observing the impact of the installation and the adoption process, we decide to approach an in-the-wild study with Selfie Cafe. We collected data from observations and spontaneous comments, without any interference from the researchers. The space chosen was a public space near to the community kitchen area of the computing department of a public university.

Before installing the application, we carried out an analysis of the space, in order to understand how people were using that space. After 2 weeks, we found that between the classes were the times people crowded that space in order to buy a coffee in the coffee machine. However, the time they spend in there were quite short, 5 minutes in average including the 2 minutes to get a coffee in the machine.

Regarding people's privacy, we placed a folder announcing generically the study to prevent influencing the results, making people aware of the study.

Selfie Cafe installation evoked interest from people who interact with it. After visualizing their image projected on the wall, people usually tried to explore more. They were curious on discovering the purpose of the installation and if it was interactive. This curiosity becomes clear in some spontaneous comments, such as, "Oh my God, this is so cool. Can I try it?" and "Is this real? I want to try it." This states the appealing factor of the installation.

After exploring the installation, people started to realize that they could use it by touching on the image projected on the wall. We observed that once someone was willing to take the photo, they usually started to laugh and they want to take the photo with someone close by. This fact can be the opportunity for interacting with a stranger.

When the photo was taken, people looked excited and curious in understanding how the installation works. This curiosity can leverage discussions among people around. Once we deployed in a computer department, the installation evoked some theories in our audience. They started to discuss on how the installation probably works and what technologies are involved in the system.

About the photos displayed on the large screen, people started to hangout around the installation commenting the photos, and we observed some of them getting their mobile device to access the Selfie Cafe website to vote. We notice some people taking a picture of their displayed picture with their mobile phones. In addition, we noticed some indices that some people were invited by others to interact with the installation.

With Selfie Cafe, we provided a social activity to create a community-meeting place where people could spend some good moment, enjoy each other's company and share conversations, laughs, and photos.

5   WishBoard

WishBoard is an explorative artistic representation of the community wishes. The installation invites people to share openly their individual expectations, thoughts, and feelings in a public space. WishBoard aims at promoting the sense of community through a collaborative artistic expression of the wishes of its participants. This kind of self-expression is a powerful sign of individual freedom in western culture, revealing people internal attributes, such as preferences, opinions, and values. Since we all have dreams, plans and goals for the future, the installation works as a reminder for people of a community to pursue their dreams, contributing to strengthen inter-connectedness among people (Ferreira et al. 2014a).

The main concept of this project is a reinterpretation of a chalk-and-wall-based art installation called Before I Die. Before I Die is an art project that invites people to reflect on the finitude of life, pick a chalk and write on a chalkboard painted wall their individual aspirations for before death in a public space (Chang 2013). On the other hand, WishBoard celebrates life giving people a chance to share their future goals, creating a social place where people may attend to discuss their wishes, socialize, and relief from the stressful demands of everyday life. In this democratic space, community social life can flourish, which can lead to stronger community ties.

WishBoard comprises a concept of using technology to share wishes in public spaces aiming to provide a sense of community. We are exploring different meta-phors to embody this project, as shown in Fig. 2.

For a community, having spaces for self-expression and socialization plays a crucial role in keeping their own culture alive (Brenny and Hu 2013). WishBoard gives people an opportunity to deeper know the community and the local people expectations for future. In addition, people can learn about the culture of the group, including its values, attitudes, and expectations, once that is part of the process of socialization (Corcoran and Clark 1984). From the perception of similarity to others, people can feel part of a larger dependable and stable structure, defined as sense of community (McMillan and Chavis 1986).

5.1   Installation Design

The inspiration for this project came from the Before I die project (Chang 2013). Aiming to provoke a similar feeling and engagement on the audience we reinter-preted it using the concept of translating art using ICT (Ferreira et al. 2014a) and the embodiment design (Fels 2004).

According to Fels (2004), people build relationships with external objects to their own self depending on how deeply embodied the person is into an object or an object is into the person. He suggests four relationships: response, control, contemplation, and belonging. These types of relationships occur during the interactive experience

Fig. 2 WishBoard installations exploring multiple screens and projections

and may overlap, increasing the intimacy relation between a person and an object. In both installations, WishBoard and Before I Die, the first relationship, called response, is exploratory and begins with the first contact with the installation. People are attracted to the installation trying to understand more about it and, depending upon if the person's expectations were satisfied, the relationship intensity increases. Regarding the second relationship, control, the person feels able to interact with the installation using either a mobile device or a chalk. The contemplation occurs when people begin a dialogue with the displays or wall, starting a reflection about the messages or initiating a guessing about whom sent those messages. The most intimate relationship, belonging, can occur when people participate of the artwork contributing with their message that becomes part of the installation, giving the sense of community belonging to the user. Furthermore, the messages and the quantity of messages can afford people to spend more time on the installation.

Aiming at building a great experience leading to an intimate relationship between audience and the installation, WishBoard considered imaginary and dreams as key metaphors. The main screen represents a "wish catcher," in reference to the "dream catcher" from Native American culture. This "wish catcher" selects the wishes giving a visible space for them. People can follow the shared wishes coming from the main screen, flying through the clouds and going into other screen. Moreover, every time a new wish comes to the installation, the system presents the message with a typographic art animation for few seconds, expressing the uniqueness of the wish. After that, the message goes to the "clouds" and joins to the collective on the wall, becoming part of the installation. With this collaborative expression, the installation provides the sense of belonging to the community and place attachment, as if a part of themselves was rooted in that place. In addition, the main screen creates an illusion of a "window" to a remote place linking that virtual place to the physical place when the message comes out of the screen and goes to the wall or other screen.

This project explores the interaction with contextualized situated public displays using personal mobile phones to build an intimacy and embodiment relationship between people and people, and people and the dynamic art-system. WishBoard offers a space for self-reflection and contemplation, creating a rich aesthetic interactive experience on the behavioral, visceral, and reflective levels.

In a process of self-reflection, using words to express thoughts can make people feel more committed and bound to them. Regarding the sharing of wishes on WishBoard, people just need connect to the installation wireless signal using their mobile device and then fill a sentence. This creates a link to the place as a space for self-expression, leveraging to place attachment and supporting the occurrence of thirdplaceness.

In order to maintain a playful mood in the installation and avoid display unsuitable sentences, WishBoard has a keyword-based filter. WishBoard system is web-based written in a combination of HTML5, CSS3, and JavaScript. In order to control and synchronize the screens we implement a websocket server.

5.2 Discussion

In order to understand the audience behavior and the impact of the technological installation in a public space, we carried out an in-the-wild study with WishBoard. We collected data from shared messages, spontaneous comments and video recordings filming both users and the installation. In addition, a researcher was present in a strategic place, taking field notes without disturbing the normal characteristics of the space. We deployed WishBoard at a hall of a university department in two different occasions, totalizing 2 weeks of experiment.

To deal with the privacy, we used the mailing list of department to inform people about the installation, as well as data collection, the presence of cameras, and privacy policies. In addition, before posting a message, the system presents to users our privacy policies and the ethical agreement on the use the collected data only for research purposes. Besides that, we did not collect personal data (e.g., name, e-mail, and age) and we de-identified all data that can identify participants or that might embarrass them.

In order to make installation visible for passersby, we used the central and peripheral model. In this model, Weiser and Brown (1997) argue that technology can engage users without generating a visual or cognitive discomfort. The model of central and peripheral attention describes that objects can attract the viewer's attention, even though the peripheral zone of vision. According to this model, much of our brain is devoted to processing peripheral (sensory). Thus, when there is something unusual in place, you can capture and bring this new information to the central area of the user's attention.

During the analysis of the recordings and the notes, we observed that users felt more involved and committed in interacting with the installation by posting a message on WishBoard when they were in a group (68 % of the messages). A behavior observed in the groups formed around the installation, was a competition for the most creative or funniest message. Moreover, it became clear that in many situations, the public around the installation progressively increased when there was a group in front of screens, forming a 'buzz' social in place and creating the effect defined as honey-pot (Brignull and Rogers 2003).

Beyond the honey-pot, sometimes people were attracted to the installation, making a late stop and walking back, trying to explore and understand more about the installation. Unlike the honey-pot, this effect, known as landing effect (Müller et al. 2012), occurred more frequently when there was no one in the installation and in some cases when people were in a hurry and through peripheral vision sensed something different in place.

During the experiment and analysis, moments found that people appropriated the installation. According to Salovaara et al. (2011) appropriation refers to the creative and innovative ways in which users adapt technologies, assigning a new purpose and adapting the solution to their goals. In WishBoard installation, people subverted the proposed installation in some situations, sending messages related to immediate present rather than the future, as the theme of the installation suggested. An example of this appropriation is in the following sentence: "For next year I want a coffee." Another example is the use of the installation as a messenger to exchange messages, sometimes going with a relaxed tone, such as the message: "For this year I want Lucas really work." In addition, people sent emoticons and common expressions in internet communication contexts, as "hauhauhau" to express laughs by Brazilians (Ferreira et al. 2015b). This suggests that, in public spaces, installations must be prepared for the honey-pot and landing effects. Besides that, self-expression tools need to allow appropriation of use, as a sign of new uses for the installation, reflecting the needs of some users.

Attracting the attention of users through interactive installations is not a quite simple task for designers. In public spaces, the challenge is even greater due to the presence of many other objects competing for user attention (Müller et al. 2012). In WishBoard installation, we observed passersby turning their heads to the installation area, feeling attracted by the public display. Some people mentioned having curiosity in testing the system, asking questions about the installation to other people on the site. In the first deployment, approximately 30 % of passersby stayed in the installation and more than half of these people remained in the place for more than 1 minute. In the second, about 25 % of passersby stayed in the installation and about 60 % of them remained in the place for more than 1 minute. Moreover, in both deployments we counted approximately 325 users sending messages in front of screens (Ferreira et al. 2015b).

To achieve the effectiveness of the installation, the attractiveness plays a key role to draw the audience's attention and engage them in using WishBoard. Since this appeal goes beyond aesthetics and encompasses factors related to the physical location of deployment, disposal of the installation and the ease of access to the site. Moreover, the effectiveness of the public display raises an issue of whether the system is able to maintain the public attracted and engaged for long term.

In both deployments with WishBoard, people appropriated of the space, previously used only as access to the department, as a meeting place to introduce to each other and talk about their common interests. People shared democratically their personal mobile devices and the furniture available. The displayed messages in the public displays leveraged discussions and even laughs. For example, a group of people began to discuss about the messages displayed on the screen attempting to guess who could have sent those messages, showing an interest in knowing more about their community. During the analysis, we found the occurrence of unexpected encounters between friends and acquaintances, as presented in Fig. 3.

The space, previously socially abandoned, transformed into a social place for students, faculty, staff and visitors to share casual and informal conversations. We

Fig. 3 User greeting an acquaintance while he was using WishBoard (Ferreira et al. 2015b)

noticed that public displays support the notion of community built and maintained by experience and interaction with and in place. In addition, self-expression and content provided by these public displays can lead to leverage discussions among people.

Attractive design systems that are able to make users feel encouraged and interested in interacting with them compose one of the biggest challenges in public installations. Nevertheless, keeping people involved in a continuous and regular use with the installation is even more difficult. Regular are important for installation that promotes thirdplaceness because they shape the "tone" of space, encouraging others to use the installation and the space. In this context, we considered as regular people who sent messages for installation at different moments (two or more times) and invited other people to interact with WishBoard. In the first deployment, we identified seven regulars in the second implantation four other regulars. To illustrate how we identified these regulars, Fig. 4 shows some of these regulars.

Regarding privacy concerns, we perceived in WishBoard people willing to show their message to others, pointing their finger to the message or even calling someone

Fig. 4 Users using WishBoard at different occasions, engaging others to use the installation and talking about the shared wishes in a socialization process (Ferreira et al. 2015b)

Fig. 5 **a** Student celebrating a message sent to her and **b** users taking pictures of the messages they sent on WishBoard (Ferreira et al. 2015b)

to see your message. For example, a user called a friend to go in front of the screen because he will send a message for her wishing that she finished her doctoral degree. After she reads the message for here, as response to the surprise, she smiled and raised his hands in celebration, as shown in Fig. 5a. In addition, some people took pictures of their messages, as presented in Fig. 5b.

In the collected comments, people said they realized that it was not possible to identify your messages, making the system use more comfortable for them. Besides that, no one expressed any concerns about privacy. For Chang (2013), in public installations anonymity play an important role allowing shy people to express themselves more easily. This anonymous nature of the wishes creates a neutral and leveling ground, where everyone can participate without worrying about their status in the community.

In our deployments, we have noticed that people kept their spirits high, avoiding spreading bad feelings. Moreover, we are thinking in engaging the community to report the unsuitable sentences to help in the maintenance of the installation.

6   Conclusions and Future Work

In this chapter, we focus on how ICT can promote thirdplaceness and support communities. For that, we designed two installations and deployed them in-the-wild to observe the occurrence of thirdplaceness in a place augmented by ICT and that is not a third place. Thirdplaceness gives people the experience of a "great good place" transforming a space in a third place for a period or permanently.

Third place is a community-meeting place where people can talk freely, openly, and entertain without caring about their social status. Oldenburg (1999) describes third places as a means to "keep in touch with reality" promoting close personal ties outside the home and workplace. These sites have a key role in community life by providing a democratic environment and available to discuss topics such as politics, sport and events in the region. The neutrality of these spaces allows people to express themselves spontaneously, thereby strengthening the sense of belonging and sense of community. However, these places are disappearing, partly because of unplanned urbanization process and the modern lifestyle, leading people to have fewer opportunities to attend such spaces. This lack of third places can affect quality of life of individuals in a community. Therefore, thirdplaceness is the "event" where and when the characteristics of a third place are reached, creating the feeling of being in a third place. Furthermore, providing opportunities for socialization of a third place and confronting the isolation stigma of using technologies in public spaces.

Based on the third places characteristics defined by Oldenburg (1999) and in our observations, we believe that to achieve thirdplaceness people need to feel free to be and express themselves. Promoting tools for self-expression can support to achieve this feeling. Moreover, providing anonymity can comfort audience, mainly shy people, giving them a chance to express themselves. People might have the same privileges and opportunities to participate in the activities, giving leveler awareness in such space. The experiences and interactions with and within the place can empower relationships among people and promote place attachment. ICTs can encourage activities among its users such as competitions, and provide information that can leverage discussions and conversations. Besides that, people had a chance to socialize taking the advantage of the Honey-pot effect. The place needs to be easy to access and allows appropriation of use to provide the feeling of fulfilled needs in the occupants. ICT installations need to ease to use and accessible providing information and, eventually, even Wi-Fi signal. Encouraging self-expression can often promote the relative feelings of warmth, possession, and belonging for people and they can feel that a piece of themselves is rooted in that space.

We demonstrate the relevance of supporting self-expression in public spaces through interactive installations. These installations create an environment to people express their thoughts, feelings, aspirations, emotions, and identity. During the deployments, people shared the environment to discuss various topics and enjoying each other's company. The installations gave people a chance to socialize and provided information that leveraged conversations reinforcing their notion of community. Our study found that allowing self-expression offers the neutral ground to people express freely their individuality. In addition, we showed the importance of allowing appropriation of use in self-expression. Overall, our research reinforces the crucial role that public displays and mobile devices can play in providing an affordable way for people to express their identity, promoting thirdplaceness.

As future work, we hope to revise the third places characteristics for the concept of thirdplaceness as Memarovic et al. (2014). Furthermore, we also hope to explore other technology interventions to support a sense of community in further different settings.

Acknowledgments We want to thank all the participants and everyone from the Advanced Interaction Laboratory at Federal University of Sao Carlos who collaborated in conducting the studies presented in this chapter. We gratefully acknowledge Boeing, FAPESP and CAPES for the partial fund.

References

Abouzied, A., Chen, J.: CommonTies: a context-aware nudge towards social interaction. In: Proceedings of the 17th ACM Conference on Computer Supported Cooperative Work & Social Computing, pp. 1–4. ACM Press (2014)

Agren, P.: Virtual community life: a disappearance to third places for social capital. In: Proceedings of the 20th Information Systems Research Seminar in Scandinavia (IRIS 20) "Social Informatics", pp. 683–694. Department of Informatics, University of Oslo, Oslo (1997)

Anacleto, J.: Culture sensitive ICT solutions to improve quality of life: supporting changes for innovation. In: Brazil-US Symposia on Future of Cities—Frontiers of Science and Engineering. Rio de Janeiro. URL http://www.nasonline.org/programs/kavli-frontiers-of-science/brazil-us-frontiers/ (2014). Accessed Apr 2016

Brenny, S., Hu, J.: Social connectedness and inclusion by digital augmentation in public spaces. In: Proceedings of the 8th International Conference on Design and Semantics of Form and Movement (DeSForM 2013), pp. 108–118. Philips, Wuxi (2013)

Brignull, H., Rogers, Y.: Enticing people to interact with large public displays in public spaces. In: Proceedings of the IFIP International Conference on Human-Computer Interaction (INTERACT 2003), pp. 17–24 (2003)

Bueno, A., Anacleto, J., Calderon, R., Fels, S., Lea, R.: ICT to support community gardening: a system to help people to connect to each other in real life. In: Proceedings of Designing Interactive Systems (DIS 2014), pp. 133–136. ACM Press (2014)

Calderon, R., Fels, S., de Oliveira, J., Anacleto, J.: Understanding NUI-supported nomadic social places in a Brazilian health care facility. In: Proceedings of the 11th Brazilian Symposium on Human Factors in Computing Systems, Brazilian Computer Society, pp. 76–84 (2012)

Chang, C.: Before I Die. St. Martins Griffin, New York (2013)

Corcoran, M., Clark, S.M.: Professional socialization and contemporary career attitudes of three faculty generations. Res. Higher Educ. **20**, 131–153 (1984)

Doheny Farina, S.: The Wired Neighborhood. Yale University Press (1998)

Farnham, S., McCarthy, J., Patel, Y., Ahuja, S., Norman, D., Hazlewood, W., Lind, J.: Measuring the impact of third place attachment on the adoption of a place-based community technology. In: Proceedings of the SIGCHI Conference on Human Factors in Computing Systems, pp. 2153–2156. ACM (2009)

Fels, S.: Designing intimate experiences. In: Proceedings of the 9th International conference on Intelligent user interfaces, pp. 2–3. ACM, New York, NY, USA (2004)

Ferreira, V., Anacleto, J., Bueno, A.: Translating art installation into ICT: lessons learned from an experience at workspace. In: Proceedings of the 32nd ACM International Conference on the Design of Communication CD-ROM, p. 11. ACM (2014a). doi:10.1145/2666216.2666226

Ferreira, V., Anacleto, J., Colnago, J., Bueno, A.: Promoting Thirdplaceness with an ICT Interactive Art Installation at the Workspace. In: Hacking HCI3P: Second Workshop on Human Computer Interaction for Third Places. ACM, Vancouver, Canada (2014b). doi:10.13140/RG.2.1.2085.3207

Ferreira, V., Anacleto, J., Bueno, A.: WishBoard: promoting self-expression in public displays to leverage the notion of community. In: Workshop: Doing CSCW Research in Latin America: Differences, Opportunities, Challenges, and Lessons Learned (CSCW' 2015), vol. 1, Vancouver, Canada (2015a). doi:10.13140/RG.2.1.3273.1687

Ferreira, V., Anacleto, J., Bueno, A.: Sharing wishes on public displays: using technology to create social places. In: Human-Computer Interaction—INTERACT 2015, pp. 578–595. Springer (2015b). doi:10.1007/978-3-319-22698-9_40

Gaved, M., Anderson, B.: The impact of local ICT initiatives on social capital and quality of life. Chimera Working Pap 6, 1–35 (2006)

Griswold, W., Shanahan, P., Brown, S., Boyer, R., Ratto, M., Shapiro, R., Truong, T.: ActiveCampus: experiments in community-oriented ubiquitous computing. Computer 37(10), 73–81 (2004)

Hardy, J.: UbiDisplays: a toolkit for the rapid creation of interactive projected displays. In: The International Symposium on Pervasive Displays (2013)

Kendall, L.: Hanging out in the virtual pub: masculinities and relationships online. University of California Press (2002)

Khermouch, G., Veronsky, F.: Third places. Brandweek 36(11), 36–40 (1995)

Kim, H., Sherman, D.: "Express yourself": culture and the effect of self-expression on choice. J. Pers. Soc. Psychol. 92, 1 (2007)

Kim, J., Lee, J., Lee, H., Paik, E.: Design and implementation of the location-based personalized social media service. In Internet and Web Applications and Services (ICIW 2010), pp. 116–121. IEEE Press (2010)

Lane, G., Thelwall, S., Angus, A., Peckett, V., West, N.: Urban Tapestries: Public Authoring. Place and Mobility. Proboscis, Southampton, UK (2005)

Marshall, P., Morris, R., Rogers, Y., Kreitmayer, S., Davies, M.: Rethinking 'multi-user': an in-the-wild study of how groups approach a walk-up-and-use tabletop interface. In: Proceedings of the SIGCHI Conference on Human Factors in Computing Systems, pp. 3033–3042. ACM (2011)

McCarthy, J., Farnham, S., Patel, Y., Ahuja, S., Norman, D., Hazlewood, W., Lind, J.: Supporting community in third places with situated social software. In: Proceedings of the Fourth International Conference on Communities and Technologies, pp. 225–234. ACM (2009)

McMillan, D., Chavis, D.: Sense of community: a definition and theory. J Community Psychol 14(1), 6–23 (1986)

Memarovic, N., Fels, S., Anacleto, J., Calderon, R., Gobbo, F., Carroll, J.M.: Rethinking third places: contemporary design with technology. J. Community Inform. Spec. Issue Urban Plann. Community Inform (2014)

Memarovic, N., Langheinrich, M., Alt, F.: The interacting places framework: conceptualizing public display applications that promote community interaction and place awareness. In: Proceedings of the 2012 International Symposium on Pervasive Displays, p. 7. ACM (2012)

Müller, J., Walter, R., Bailly, G., Nischt, M., Alt, F.: Looking glass: a field study on noticing interactivity of a shop window. In: Proceedings of the SIGCHI Conference on Human Factors in Computing Systems, pp. 297–306. ACM (2012)

Oldenburg, R.: The Great Good Place: Cafes, Coffee Shops, Bookstores, Bars, Hair Salons and the Other Hangouts at the Heart of a Community. Marlowe & Company, New York (1999)

Pasick, A.: More than love to be found on networking sites, USA Today. http://usatoday30.usatoday.com/tech/webguide/internetlife/2004-02-25-online-networking_x.htm (2004)

Paulos, E., Goodman, E.: The familiar stranger: anxiety, comfort, and play in public places. In: Proceedings of the SIGCHI conference on Human factors in computing systems, pp. 223–230. ACM Press (2004)

Peltonen, P., Kurvinen, E., Salovaara, A., Jacucci, G., Ilmonen, T., Evans, J., Saarikko, P.: It's Mine, Don't Touch!: interactions at a large multi-touch display in a city centre. In: Proceedings of the Conference on Human Factors in Computing Systems, pp. 1285–1294. ACM Press (2008)

Preece, J., Rogers, Y., Sharp, H.: Interaction Design: Beyond Human-Computer Interaction. Wiley, New York (2002)

Putnam, R.: Bowling Alone: The Collapse and Revival of American Community. Simon & Schuster, New York (2000)

Rogers, Y., Connelly, K., Tedesco, L., Hazlewood, W., Kurtz, A., Hall, R.E., Toscos, T.: Why it's Worth the Hassle: the Value of In-Situ Studies When Designing Ubicomp, pp. 336–353. Springer, Berlin Heidelberg (2007)

Rogers, Y.: Interaction design gone wild: striving for wild theory. Interactions 18(4), 58–62 (2011)

Salovaara, A., Höök, K., Cheverst, K., Twidale, M., Chalmers, M., Sas, C.: Appropriation and creative use: linking user studies and design. In: CHI'11 Extended Abstracts on Human Factors in Computing Systems, pp. 37–40. ACM (2011)

Salvador, T., Sherry, J., Urrutia, A.: Less cyber, more café: enhancing existing small businesses across the digital divide with ICTs. Inf. Technol. Dev. 11(1), 77–95 (2005)

Sarason, S.: The emergence of a conceptual center. J. Community Psychol. 14(4), 405–407 (1986)

Scheible, J., Tuulos, V., Ojala, T.: Story Mashup: design and evaluate on of novel interactive storytelling game for mobile and web users. In: Proceedings of the 6th International Conference on Mobile and Ubiquitous Multimedia, pp. 139–148. ACM Press (2007)

Schuler, D.: New Community Networks: Wired for Change. ACM Press/Addison-Wesley Publishing Co. (1994)

Shipley, J.: Selfie love: public lives in an era of celebrity pleasure, violence, and social media. Am. Anthropol. 117(2), 403–413 (2015)

Silva, A., Hjorth, L.: Playful urban spaces: a historical approach to mobile games. Simul. Gaming 40(5), 602–625 (2009)

Silva, R., Anacleto, J.: Providing ICT support to promote communities' emotional balance. In: Proceedings of the 7th International Conference Social Computing and Social Media 2015, pp. 78–88. Springer (2015)

Silva, R., Colnago, J., Anacleto, J.: Design de aplicações para interação em espaços públicos: formalizando as lições aprendidas. In: Proceedings of the 13th Brazilian Symposium on Human Factors in Computing Systems, pp. 120–129. Sociedade Brasileira de Computação (2014)

Soukup, C.: Oldenburg's great good places on the world wide web computer-mediated communication as a virtual third place. New Media Soc. 8(3), 421–440 (2006)

Turkle, S.: Virtuality and its discontents: searching for community in cyberspace. Am. Prospect 24, 50–57 (1996)

Walker, R.: Crossfire. The New York Times Magazine. http://query.nytimes.com/gst/fullpage.ht ml?res=9506E0D81E38F93BA15750C0A9669D8B63 (2010). Accessed March 2016

Weiser, M., Brown, J.: The coming age of calm technology. In: Beyond Calculation, pp. 75–85. Springer, New York (1997)

Wellman, B.: The privatization of community: from public groups to unbounded networks. In: Abu-Lughod, J. (ed.) Millennial milestone: a "switching crisis" in sociology, pp. 89–104. International Sociological Association, Barcelona (1998)

Zheng, Y., Capra, L., Wolfson, O., Yang, H.: Urban computing: concepts, methodologies, and applications. ACM Trans. Intell. Syst. Technol (TIST) 5(3), 38 (2014)

Zheng, Y., Liu, Y., Yuan, J., Xie, X.: Urban computing with taxicabs. In: Proceedings of the 13th International Conference on Ubiquitous Computing, pp. 89–98. ACM (2011)

Mischief Humor in Smart and Playable Cities

Anton Nijholt

Abstract In smart cities we can expect to witness human behavior that is not necessarily different from human behavior in present-day cities. There will also be demonstrations, flash mobs, urban games, and even organized events to provoke the smart city establishment. Smart cities have sensors and actuators that maybe can be accessed by makers and civic hackers. Smart cities can also offer their data to civic hackers, who may create useful applications for city dwellers. Smart cities will have bugs that can be exploited for fun or appropriation. Humor is an important aspect of our daily activities and experiences. In this chapter, we explore how humor can become part of smart and playable cities. We do this by investigating the role of humor in game environments. In games, we have accidental humor, for example because of bugs, and we have humor that occurs because a gamer wants it to happen. This latter type of humor can be produced by looking for bugs, by not following the rules of the game, or by intentionally creating situations that lead to humorous events in the game. This may certainly include humor at the expense of others. We investigate how such views of game humor can find analogs in the humor that may appear and be created in smart and playable cities.

Keywords Playable cities · Computational humor · Accidental humor · Hacking · Trolling · Griefing · Bullying · Bugs · Smart cities · Games

1 Introduction

In this chapter, we investigate possible occurrences of humorous events in smart environments such as playable cities. The concept of playable cities was introduced

A. Nijholt (✉)
Imagineering Institute, Medini Iskandar, 79200 Iskandar Puteri, Johor, Malaysia
e-mail: anton@imagineeringinstitute.org; a.nijholt@utwente.nl

A. Nijholt
Faculty EEMCS, Human Media Interaction, University of Twente, PO Box 217
7500 AE Enschede, The Netherlands

© Springer Science+Business Media Singapore 2017
A. Nijholt (ed.), *Playable Cities*, Gaming Media and Social Effects,
DOI 10.1007/978-981-10-1962-3_11

in Bristol (UK). Through various projects, citizens have been able to participate in funny events made possible by sensor and actuator technology. In other cities, sensors and actuators have been embedded in street furniture such as traffic lights, mailboxes, lamp posts, escalators, shop windows, and billboards (prankvertising). Humorous events can be staged; they can occur spontaneously; they can occur accidentally; and they can occur through nontraditional explorations of the environment, such as by looking for bugs or unusual situations that have not been foreseen by the designer and that invite or provoke humor. However, smart worlds are digitally enhanced physical worlds, worlds that are designed using sensors and actuators that can be controlled and adapted by designers, owners, or inhabitants of these environments. Can this be done in a way that encourages the emergence of humorous situations?

In artificial worlds, we can design humor. In a 'language world,' we use words, speech, prosody, and timing to tell jokes. In cartoons we integrate drawings and text. In animated movies, we are free to play with the laws of gravity and can recover in no time from the worst of injuries. In comedy and movies, the stage manager or film director can direct the actors and events in such a way that humorous situations can occur. Humor theory offers us the notions of incongruity and incongruity resolution to help us analyze, understand, and generate humorous situations in language, cartoons, animation, comedy, movies, and the physical world. To see how we can facilitate the occurrence of humor in smart environments, it is useful to consider humor techniques as they are used in language, comedy, movies, et cetera to see how these techniques can generate humor in smart environments, where (in addition to humor that occurs spontaneously) we also have the opportunity to 'stage' humorous events by introducing incongruities using sensors and actuators.

It is particularly useful to look at humor in videogame environments. Gamers are confronted with 'canned' humor. This canned humor can be integrated into the game and is triggered by actions performed by a gamer. In a way that is usually very limited, some context and history can be included in a generated humorous utterance or event that fits the narrative. Clearly, in MMORPGs teams can also discuss and introduce strategies that aim at cultivating team behavior and exploiting the game environment in such a way that a sequence of events will lead to a hilarious situation (usually including the humiliation of their opponents). In games there is not always control over the consequences of gamers' actions. Not every action of a gamer can be anticipated. This can lead to accidental humor, for example, when a collision detection algorithm is not always perfect. Creative play with a game engine or misuse of a game engine is also possible. This may include hacking the game engine to create mischief humor. Some game hackers and digital game mischief makers may also disregard the aims of the game and the game's narrative.

Herein, we argue that game environments and digitally enhanced real worlds, such as digital and playable cities will converge. In Sect. 2 we briefly review the theories and techniques of humor. Usually, theories and techniques focus on language humor (jokes, wordplay, and conversational humor). Moreover, in comedy, in (cartoon) movies and also in games humorous situations are created using exaggerations and behavior that are not always possible in the real world. Section 3 is about game environments. The design of humor in a game needs to be carefully executed. As in the real world, the introduction of humor needs to satisfy some appropriateness criteria. There can be humor in interaction with game characters, and, just as in the real world, there can be accidental humor. Accidental humor can occur because of shortcomings in the game technology, leading to unexpected and not always gamer controlled situations. Humor can also emerge when teams of players try to trick each other. In Sect. 4 we discuss some activities in game communities that can be considered pranking, mischief or malicious behavior, including cheating and hacking of the game mechanics. Such behavior can be fun and is often used to make fun of other—not so clever—gamers. In Sect. 5 we make the transition from game worlds to digitally enhanced real worlds, such as digital and playable cities. We provide some examples of humor using real-world sensors and actuators, and we offer observations on how the various types of humor that we distinguish in digital game environments can appear and can be controlled in smart and playable digital cities. Some conclusions follow in Sect. 6.

2   On Humor: An Introduction

In this section we briefly review the various theories of humor. In the sections that follow we mainly look at the so-called incongruity or incongruity resolution theory. This theory emphasizes the cognitive shift we must make when our first interpretation of a situation has to be replaced by a second, and this second interpretation makes clear what we wrongly understood in our first interpretation of the situation. When these interpretations are truly in contrast with each other, humor results. From the point of view of generating humor in game or real-world environments, this incongruity humor viewpoint demands that we design either surprising events or situations where surprising events can happen. Sensors in smart environments allow us to detect, recognize, and interpret human behavior. Actuators and computer intelligence allow us to provide humorous feedback, including making changes to the environment that lead to (potentially) humorous situations. Sensors and actuators allow us—and the environment—to create ambiguous situations that allow multiple interpretations and to create unexpected and surprising events.

Theories of humor focus on verbal humor. We (very) briefly review these theories. In humor research a distinction is often made between the cognitive aspects of humor, the function of humor and the effect of humor. The cognitive aspects address unexpectedness, surprise, incongruity, and resolving or understanding incongruity. This view of humor is known as the *incongruity* or *incongruity-resolution theory*, and it usually requires a cognitive shift. We interpret a particular situation and do not expect anything unusual, but we are then confronted with new information that requires us to reinterpret the situation. When these interpretations are sufficiently opposed, humor results. It has been argued that humor always requires some incongruity. In language this incongruity usually appears sequentially, requiring a shift in perspective in time. In visual or audiovisual humor (or humor that involves other modalities) incongruities can appear at the same time; we have a simultaneous interplay of perspectives. However, most research on incongruity theory addresses language humor and the simultaneous display of incongruities is not part of these studies. However, it should be part of the discussion when we discuss nonverbal humor in physical or virtual environments. In 1779, James Beatty was already talking about "two or more inconsistent, unsuitable, or incongruous parts or circumstances, considered as united in one complex object or assemblage, as acquiring a sort of mutual relation from the peculiar manner in which the mind takes notice of them" Kant, Schopenhauer and Kierkegaard are among the other philosophers who paid attention to incongruous humor.

As mentioned above, most existing theories of humor apply to modeling verbal humor. These theories can be found in humor textbooks (Raskin 2008). The second theory we want to mention is the *theory of superiority or disparagement*, which is linked to names such as Plato, Aristotle and Hobbes. It assumes that we laugh at the misfortune or inferior position of others. Slipping on a banana peel is an example. However, mostly we experience it in verbal jokes. For example, *"How do you make a blonde laugh on Saturday? Tell her a joke on Wednesday."*

A third theory of humor is associated with Sigmund Freud and is called the *relief theory*. Freud describes humor as a necessary means of releasing pent up frustration originating in unpleasant experiences or social and sexual taboos (Freud 1905). Freud cites the following joke as an example. *A royal personage was making a tour through his provinces and noticed a man in the crowd who bore a striking resemblance to his own exalted person. He beckoned to him and asked*: *"Was your mother at one time in service in the Palace?" "No, your Highness"* was the reply, *"but my father was."* AI expert and philosopher Marvin Minsky built on this by mentioning cognitive taboos that are breached when jokes defy logic (Minsky 1981). For example, *"Ethel orders a pizza. The waitress asks her whether she would like it cut into four or eight slices. Ethel answers 'Just four, I'm on a diet.'"*

These theories emphasize different functions of humor. Superiority theory addresses the social aspects of humor. We observe or are told about a person or a situation where we would want to behave differently or be treated differently from the protagonist. We do not want to be the person who is slipping on a banana peel, and we do not think such a thing will happen to us. Often a joke makes us laugh because of someone's stupid behavior or because of behavior that is not in agreement with the professional or moral behavior we generally expect. The relief theory point of view addresses emotions, particularly the relaxation of tension. This may concern the teller of a joke or the creator of a humorous situation, as well as the listener and observer who experience the humor that is created. Incongruity theory tries to identify the cognitive mechanisms of humor. How do we experience unexpected events? How do we address new perspectives and potentially ambiguous interpretations of events, whether in language or in the real world? Incongruity theory emphasizes the stimuli that produce humor. When considering how to introduce humor into game or smart environments, including smart and playable cities, we have sensors and actuators that can be employed to design stimuli that may lead to humorous events (Nijholt 2014, 2015a, b) or that can help mischief makers to create humorous events in smart environments. For that reason, in the remainder of this chapter we will focus on humorous events and behaviors that can be created using (virtual) sensors and (virtual) actuators in virtual, mixed reality and real worlds. Obviously, embedding smart technology in our 'real' world makes our real-world properties of virtual worlds, in which there are many more opportunities to control events and users or inhabitants participating in these events.

The concept of 'Computational Humor' was introduced in 1996 with the organization of a conference on computational humor at the University of Twente (the Netherlands) (Hulstijn and Nijholt 1996). The focus was on verbal humor, and in recent research the focus is still on verbal humor. This is understandable because research in linguistics, and in particular computational linguistics, looks at formal properties of language and dialogue. Among the issues being investigated are nonliteral language use, ambiguities in language use and, to a lesser extent, irony and sarcasm. Recent research on computational humor often involves machine learning algorithms that use 'big linguistic data,' for example, sentences and texts collected from the Worldwide Web (Mihalcea 2007). This research does not address physical humor, humor that involves physical behavior, or events that take place in a physical environment. Currently, it also does not address events that take place in digitally enhanced physical environments. There exist, however, typologies of humor that do include physical humor (Morreal 1983; Berger 1993; Buijzen and Valkenburg 2004). These typologies are useful in characterizing humorous events generated by gamers in game environments, and they can inspire game designers to introduce scripted humor in their games.

3 Humor Tracks in Game Environments

Game designers must make intentional decisions about including or not including humor in their games. Humor can be included in cut scenes and thus not really included in the flow of the game. However, humor can also be included in game play through sounds, music, and language. In addition to a sound track, there can be a humor track included in a videogame. Such a track must know the history of interactions that have occurred during game play; it needs to learn about a gamer's knowledge and preferences; it needs to know about possible humorous interruptions or continuations of a game situation; and it needs to know about possible humorous interactions. Despite all the research into artificial intelligence and affective computing, game environments and their non-playing characters have only very limited 'intelligence' and can only make limited and preprogrammed assessments of situations and decisions about how to continue. Apart from being able, using the elements that are available in a particular game situation, to generate humor or create a potentially humorous situation, there is of course also a decision to be made about whether it is appropriate to do so.

The design of humor in games has been discussed by psychology and human–computer interaction researchers (Dormann and Biddle 2009) as well as by game designers. We discuss mischief humor in games in the next section. Obviously, mischief humor is not about humor that has been designed to be embedded in a game or an entertainment application. Rather, it explores how to surprise, tease or even annoy people by performing unexpected and surprising activities.

There is growing interest in the possible role of humor in games. In Dormann (2014) a distinction is made between game-to-player humor, player-to-player humor, and player-to-game humor. In the game-to-player trajectory the emphasis is on scripted humor. Although the humor is scripted, some context awareness (including current game play history) can nevertheless help create variations in the humor that is generated. The player-to-player humor trajectory is about spontaneous humor in online multiplayer games, in which players use a meta-channel to discuss game events or where gamers share the same physical location. In the player-to-game trajectory, humor is generated by the gamers in the game world. This can happen accidentally or deliberately. In Dormann (2014) it is called emergent humor.

In the next section on mischief humor, we discuss the various forms of emergent humor that can be intentionally invoked by a gamer or a team of gamers. In this chapter, we do not discuss scripted humor or player-to-player humor. This does not mean that we think these forms of game humor cannot have their analogs in smart or playable cities. Accidental humor is usually regarded as humorous events that occur because of bugs in the game software. Bugs are almost unavoidable in complex software, particularly in software that is accessible to gamers and tinkerers. We cannot expect that all actions by gamers, who may not necessarily follow the rules of the game, may disregard the narrative, and may not be interested in the rewards that can be obtained, have been anticipated by the game and game mechanics designers. Mischief makers are exploiting vulnerabilities in game design. They 'screw around' and try to generate funny events, often recording them to show to other gamers.

4 Mischief Humor and Games

As mentioned in the previous section, humorous events and interactions can be designed and integrated into a game in such a way that we can talk about a humorous videogame. Unfortunately, spontaneous humorous interaction between a gamer and an NPC that requires real-time interpretation of a specific situation, including understanding the history and the context of the interaction, cannot be expected given the current state of artificial intelligence research. For this reason, it is easier to introduce humor by giving characters an unusual appearance or unusual and difficult-to-control physical behavior. We can give a character the ability to 'see' what is happening behind its back, or we can have characters that become vulnerable when they make eye contact, or we can have characters that have other nonhuman characteristics that can lead to humorous behavior. In some games, we can introduce characters that display slapstick behavior. Accidental humor appears when a gamer or its role-playing character unknowingly enters a game situation that has not been anticipated by the game designer. This is not unusual. We cannot expect a game designer to anticipate all possible actions of a gamer. This is not different from real life. For example, bridges are designed to allow traffic to travel from one side of the river to the other. Nevertheless, bridges can be destroyed, despite whatever safety coefficients have been introduced by the designer. There is a trade-off between safety and economy, where 'economy' also includes a company's attitude toward making its games hack-free.

Mischievous behavior can be a social skill in a computer game. In Sim's 4, a life simulation game, it is possible to reach various levels of Mischief Skill. Mischief behavior can lead to hilarious animations. Pranks vary from kicking over trash cans, to clogging a drain in a sink or tub, to social interactions such as asking the due date of someone who is not pregnant, or convincing someone to streak or slap a conversational partner (avatar) in the face. However, this mischievous behavior is scripted and fully embedded in the game. We are more interested in gamers' activities that allow them to prank other gamers, play tricks on other gamers, cheat during game play, or be a spoilsport. Moreover, we want to investigate how gamers amuse themselves by modifying a game, exploring bugs, hacking games, and teaming up to distort game play and the enjoyment of others. All such activities are reported in forums where gamers discuss games, strategies, cheats, hacks, and game modifications, and where their cleverness (and sense of humor) is demonstrated in 'walkthroughs' (video clips that show strange character behavior resulting from bugs or loop holes in the game mechanics) or by instructions on how to cheat, hack or modify the software.

The underlying assumption of this investigation is that we can expect similar behavior from the inhabitants of the smart and playable cities of the future.

4.1 Humor While Exploring Game Environments

In games things can go wrong. That is, in a game we can encounter a situation where the game environment is unable to react in a way that suits the aims of the game and the gamer. We have entered a situation that was not foreseen by the designer of the game. It is also possible, and gamers have adapted to this way of behavior, to search for situations where games go wrong. Can we, forgetting about the aims of the game, find game situations that lead to humorous events? Clearly, this is not about accidental humor but rather about exploring the game environment in such a way that unexpected situations will happen and preferably in such a way that we can laugh or smile about it. During this exploration, we can encounter bugs in the design of the game, we can find weaknesses, and we can find unforeseen ways of communicating with other gamers. Games have glitches. Discovering a glitch that leaves Lara Croft topless is an achievement that has to be shared with other gamers and will be rewarded with many smiles. Incongruities represented by sight gags can occur. Gamers have introduced a new genre of cinema, the genre of videogame movies called 'Machinima.' With in-game editor tools, a gamer's actions can be recorded and edited so that exploring a game environment in search of humorous situations can become part of a narrative that underlies a video that can be presented to others. This 'Machinema' genre has been discussed in Švelch (2014). Exploring a virtual environment for possibilities of humor can be called 'mischief humor.' As mentioned by Švelch (2014), this behavior can be compared with the mischief humor introduced in what was likely the first comedy movie, L'Arroseur Arrosé of 1895. In this very short movie we see a boy stepping on a garden hose and cutting of the water flow. When the gardener inspects the nozzle the boy releases the hose and the gardener gets sprayed. Rather than exploring the possibilities of a garden hose in video games, gamers can explore digital technology and have fun when they find a way to fool the game environment. Usually, this is done by disregarding the narrative and the aims of the game. In his study of Machinema humor, Švelch (2014) was able to distinguish incongruity, coincidence, slapstick, and nonsense humor as the main categories of mischief humor. "Humorous walkthroughs" is a genre in which a gamer shows how a game environment can be explored in a humorous way and where his or her comments on the game adds to the humorous effect.

4.2   Humor at the Expense of Other Gamers

Exploring a game environment to create humorous events and to collect these events in a video presentation does not necessarily require cooperation or interaction with like-minded gamers. In multi-player games there is the opportunity to create humor at the expense of other gamers. Players can team up to create an unexpected and humorous situation in which they can defeat their opponents. Communities of players will develop their own particular senses of humor and make other teams the victims of their humor. Obviously, as in the real world, gamers can find fun in pranking and bullying. Here, they are helped by digital technology and by the anonymity the Internet offers. Some games, such as the aforementioned Sims 4, offer the possibility of offending or even slapping another player.

In the Internet world and in multiplayer videogames in particular, to play a hoax on someone is also called trolling. This can be done in a friendly way, not intending any harm and making the unsuspecting victim laugh when he discovers he has been deceived. One well-known troll tactic is the 'rick roll' meme, wherein someone is led to believe that a certain action, for example, clicking a particular hyperlink, will be relevant to his aims but instead the victim is unintentionally directed to a music video for the 1987 song "Never Gonna Give You Up" by Rick Astley. When one is a member of a game community where such trolling is not only accepted but also part of communication, trolling becomes part of the game and can develop into an art. With clever trolls, initial confusion or irritation later becomes appreciation and laughter. In Fig. 1 (left) we see a troll's face on a vertically held smartphone. Not being satisfied with its orientation, we turn the device and are being trolled because the orientation changes, but not in the way we expect.

Fig. 1 Being trolled by a troll face on a smartphone

There is no clear-cut definition of trolling. Trolling can be fun and it may lead to hilarious events. In its innocent form, it should probably be called pranking. However, it can become annoying or turn into bullying and can disrupt a person's game or other Internet activity; it can even turn into digital vandalism. Trolling can include racism or sexism, and it can be done to provoke, to enact revenge or to ruin someone's game. Obviously, all this can be done in a humorous way and, if not for the victim, it can certainly be amusing for the troll or pranker and for the witnesses of the prank or the trolling.

4.3 Taking Pleasure from Grief Playing

In one game community, trolling can be considered an art, while in another community it is considered harassment, flaming or cyberbullying. Clearly, without the anonymity the Internet provides, trolls would be less popular. Various scientific papers have focused on the negative aspects of trolling (Thacker and Griffiths 2012; Buckels et al. 2014). The latter paper concludes that "… cyber-trolling appears to be an Internet manifestation of everyday sadism." This is not a view that is supported by gamers in general; however, griefing or grief play may be a different matter.

Massively Multiplayer Online Games (MMOGs) can allow players to cooperate in guilds or clans that act as communities with their own 'religions' (moral codes) and rules of conduct. Guild members are friendly to each other, but not necessarily friendly to members of other guilds. Clearly, tricking opponents is allowed, which supports spoilsports, or persons who spoil the pleasure of others. This can take rather extreme forms. Griefing is the act of purposely annoying other players, including team members. Griefers want to cause distress to other players, and they take pleasure in despicable and antisocial behavior and in sabotaging game play (Dibbell 2008; Grönroos 2013). Griefing is well known in multiplayer games such as Counter-Strike, Team Fortress, Minecraft, and EVE Online. Well-known examples of griefing are the spamming of game chart areas, blocking players, killing your own team members, infecting players with diseases and causing a plague, destroying work created by other players (Minecraft), and 'camping', or repeatedly killing the same player by waiting for him to resurrect. In Second Life, examples of griefing include invading houses and disturbing press conferences, (virtual) funerals or other activities. Griefing may inflict emotional damage on players, and inflicting such damage is one of the aims of grievers and increases their fun. Newbies and lower level characters can be easy prey for griefers, who can also team up and form 'gangs.' Ethical questions related to griefing are discussed in Warner and Raiter (2005).

4.4 Humor Created by Cheating, Hacking, and Modifying

Game cheating and game hacking communities are well known and respected communities in the world of video gamers. Cheat codes and game enhancement codes are usually shared among gamers (Consalvo 2007). Hacking a game can mean finding ways to modify game files during gameplay to manipulate game event decisions and be more successful in the game. However, hacking is also done for the fun of finding weaknesses in a design. Sometimes hackers collaborate in hacking teams and define weekly challenges to concentrate their efforts. Using hacks in online multiplayer games to obtain advantages over other players is usually disapproved of, but cheaters may use hacks or bots to win a game. Making changes to a game that affects other and future users is usually also disapproved of by gamers. Cheating and hacking do happen and are, obviously, considered problems by game designers and game companies. Categories of cheats and how to prevent them are discussed in Pritchard (2000). A multiplayer game can be ruined when there are many cheaters and their cheats are propagated.

Another way to adapt and change game environments are mods, or modifications of a video game. These can be add-ons to a game, they can replace content, or they can implement a total conversion in which only the original game engine or a modified game engine survives. Modding (Unger 2012) is part of game culture. Mods can be created by gamers and distributed using the Internet. Game companies sometimes provide mod-making tools to assist mod makers. A special category of mods are art mods; they modify games into humorous or performance art versions. These mods are also introduced with the aim of creating Machinema videos.

4.5   Humor Emerging from Controlled and Autonomous Agent Behavior

In game research, there are attempts to imbue virtual nonplaying characters with intelligence, emotions, and autonomy. That is, these attempts aim to make it possible for such characters to assess situations and act in human-like ways. Research on intelligent agents, emotional agents, and embodied conversational agents is becoming part of game research. We also meet these agents in smart urban environments. They assist us with our mobile and other wearable devices and they appear on public displays, in tangibles, as holograms or as physical social robots. Humorous interactions and humorous cooperation is possible with devices that have human-like characteristics (appearance, intelligence, emotions, sense of humor). As mentioned above, human behavior can be predicted, anticipated and, for example, using persuasion, controlled. In (serious) game and virtual reality research, we see attempts to give a 'director' role to a mediator who is able to assign roles to participants, who can guide their actions and who can introduce new events into the environment and make changes in the narrative. However, such a role can also be given to a user or gamer interested in creating his or her own game narrative and who, for that reason, needs to make changes to the environment and the behavior of its inhabitants, whether they are human or artificial.

We briefly survey some research in game environments that addresses modeling, action planning and reasoning for agents in smart environments, where planning and reasoning aim to create humorous situations. And, moreover, as discussed in this literature, it should be possible to give a human player or someone monitoring the game some responsibility for guiding others into preferred behavior and involvement in activities. As mentioned in Cavazza et al. (2003), when we want interesting behavior, we need planning mechanisms and models that do not necessarily aim at rational and optimal problem-solving behavior. A smart environment can try to understand and affect human behavior using such models, for example, with the aim of creating humorous situations. Similarly, a smart environment can use such models to affect or direct the behavior of 'autonomous' agents, tangibles, and other devices that inhabit the smart environment, whether they are human or artificial. And of course, tools based on such models and mechanisms can be used by smart city dwellers, including pranksters and mischief makers, to control and personalize their part of the smart world. This may include cheating and hacking.

Cavazza et al. (2003) introduced planning mechanisms that allow agents to continue following their aims even if certain preconditions are not fulfilled. Their research attempted to visualize the failure of the continuation of regular behavior or of adherence to the narrative in the hope that a corresponding animation of the situation becomes comical. Hence, failing plans need to be considered as dramatic mechanisms. As human residents of the smart city, we can participate, initiate and 'just' be observers of humor created by activities that fail. In this paper, heuristic search planning (HSP) techniques are used to 'control' the characters in a 'Pink Panther' script. Script writing, narrative control and role authoring are issues that need to be considered when embedding action failure and its dramatic visualization in a game narrative or in a smart environment narrative in which we want to include humor. As mentioned earlier, a comic act of (action) failing not only addresses superiority theory, it usually also includes aspects of incongruity theory.

Carvalho et al. (2012) also look at modeling the behavior of agents that act in a storytelling context, but in addition to action failures, they look at incongruities that emerge with the expected (predetermined) behavior and personality of an agent. In their architecture, personality aspects follow the well-known OCC model for emotions. This allows them to introduce characters into a narrative who behave differently from regular characters. Again, as in the previously mentioned paper, this research provides us with handles to use in creating humorous situations in smart environments. We can model the behavior of artificial agents (representing smart environments, intelligent displays, tangibles, social robots, virtual agents, et cetera) to generate humor or potentially humorous situations. Film culture and digital media center But with the help of these models, we can also embed real-time human behavior in these models in which the humans are agents in the smart environment and the models help to predict and anticipate human behavior and embed it in the smart environment. In keeping with the aim of this chapter, the models also help to introduce humor or potentially humorous situations in smart environments.

A third example we want to mention here is the research reported in Olsen and Mateas (2009). In this research, the game environment resembles a Wile E Coyote and Road Runner cartoon. Thus, we have characters (the Coyote and the Road Runner) with particular goals and a game engine that is fed by a planning mechanism to make decisions about the characters' actions. Usually, Coyote needs objects to reach his goal (to catch Road Runner). 'Gag plans' can interfere with Coyote's plans. The player or gamer in this game environment can direct the story by the manipulation of objects in this world. For example, he or she can decide to make certain objects with variable attributes available to Coyote. Entering a 'gag' plan with, for example, a rocket that will explode, will result in a failure of the original plan (to catch Road Runner).

These research examples illustrate how future smart environments can model and guide the behavior of artificial agents and their human partners in smart environments. The smart environment can offer opportunities that seem attractive and are expected to suit the goals of a human participant but will nevertheless cause the failure of his or her plans. This may lead to a humorous event. It may also be the case that someone (player, gamer, hacker, mischief maker) has control of the environment and can make changes to it to create humorous events in which human and artificial agents are involved because we can predict or model their behavior.

5 From Game Environments to Smart and Playable Cities

All the world's a stage,
And all the men and women merely players;
As you Like It, William Shakespeare, 1599

In de Lange (2015) examples are presented of serious games that allow city dwellers to participate in urban planning and design. Other urban games that are discussed provide people with an urban experience, making them aware of the environment, stimulating social interaction and inducing urban connectedness. Presently, we also see videogame worlds (e.g., Quake, Pacman, Space Invaders) that are mapped on the real world or on mixed-reality worlds. Urban games are designed in such a way that we have a game narrative in the real world (with its sensors, actuators, smart mobile devices) that usually involves some exploration of an urban environment. In this way, a game or game-like engine becomes embedded in the real world. We can take a more general view. Humans have their routines and preferences. Their behavior can be predicted and anticipated. Smart technology can also be used to persuade humans to act in a particular way and to change their behavior. So, from the point of view of a smart environment, it can 'control' a human inhabitant, just as a game engine controls a gamer by offering him or her certain choices and guiding the gamer along the possible game tracks. This happens in urban games, but it certainly is not yet the case, as we explored in Sect. 4.5, that these games exploit models of human behavior that can help to control, guide or predict actions to create humorous situations. But, as is the case with games, smart cities, or more generally, smart environments, also rely on complex software. This software contains bugs, and it can allow mischief makers to create incongruous situations by doing unusual things. It can simply allow civic hackers using publicly released data to create legal applications, in addition to built-in humor generators, that aim at bringing enjoyment to other users. Clearly, smart cities are vulnerable to criminal attacks too. Cerrudo (2015) provides an overview of the cyber security problems and possible cyber-attacks that threaten smart cities. Of course, this vulnerability can be exploited by hackers who have humorous, rather than criminal or terrorist, intentions.

5.1 The Smart City as a Stage

Smart environments are not only inhabited by humans but also by social robots, digital pets and virtual agents. This allows the environment to create situations in which humorous interactions between these human and artificial agents can emerge. However, when sensors and actuators can be manipulated or controlled by the 'gamers' among the inhabitants of smart environments, then they can also try to introduce humorous events. The smart world is a stage for mischief humor makers.

Mischief humor makers can use their access to smart urban environments in a way that is similar to that surveyed in the previous section. That is, someone can

Fig. 2 Passers-by in Bristol communicating with a mail box

explore a smart urban environment with the aim of seeing where things go wrong and how this may lead to funny situations. However, of course, pranksters will try to make fun of others using digital technology, and we can expect that cyber-bullying and trolling will not only happen in social media and virtual game environments but will also be explored using sensors and actuators in the digitally enhanced world. Cheating, hacking, and modifying are other activities that will be exported from game environments to smart urban environments.

At this moment, we see several cities introducing playful applications of digital (Internet) technology in their streets and public spaces. Citizens have access to this technology and can play with it. Because these applications are experiments and have not yet been fully integrated into a network of things, there are not yet examples of pranking, trolling, cheating or hacking in smart urban environments. Some examples of playful digital technology applied in a city environment should be mentioned. For example, in Bristol (UK) the 'Hello Lamp Post' project was introduced in 2013 under the slogan: "Bristol street objects are waking up and want to talk with you." The project allowed citizens to exchange text messages with lamp posts, mail boxes and other street furniture that had some knowledge of the environment and memories film culture and digital media center consisting of previous exchanges with passers-by (Fig. 2) that could be consulted and used in future communications.

A more recent project (2014), also realized in Bristol, is the 'Shadowing' project. In this project, lamp posts are equipped with sensors that can capture the shadows of a passer-by and can reproduce these moving shadows when someone else passes the same lamp post. Clearly, when people recognize what the lamp post is doing they start to play, introducing strange shadows and playing along with someone else's shadows (see Fig. 3).

A 2015 interactive installation, suggested by the Happy City Lab (Geneva), is an LED-sensored bench that attracts people's attention, encourages movements, attracts other potential sitters and gets them to interact (Fig. 4). More examples of introducing playful and humorous applications of digital technology can be found in Nijholt (2015b).

Fig. 3 Two examples of a lamp post casting shadows from its memory

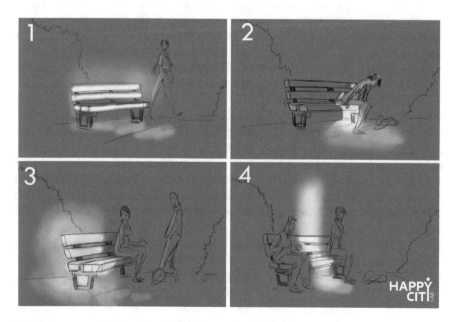

Fig. 4 Sensor-covered bench inviting people to sit and interact

5.2   *Mischief Humor in Smart and Playable Cities*

We know that smart cities will have bugs.
Anthony M. Townsend, Smart Cities (Townsend 2013, 298)

Smart and playable cities require complex software. In Cerrudo (2015), we find a list of technologies that are used to make cities smarter. They range from smart traffic control, to smart public transportation, to smart waste management to security issues. Unfortunately, this list focuses on traditional issues and possible threats to conventional smart city issues. We should look beyond this list at smart city dwellers and civic hackers who are interested in playful and humorous applications in smart cities, that is, applications that make smart cities playful and playable and allow or even invite the organization of unusual and creative events.

Whether or not the smart city wants to allow it, there will be opportunities for mischief humor. There will be bugs that can be exploited, trolls can be introduced, and cyber bullying will also be possible. Civic hackers can add humor to applications they develop. In fact, all possibilities for mischief humor that were mentioned in Sect. 4 can be introduced in smart city environments as well. Moreover, end-users can be expected to have the opportunity to modify and customize their environment (Callaghan 2007); therefore, they will also have the opportunity to allow humorous events to occur in their environment.

We cannot expect that all sensors and actuators in a smart city environment will be easily available to mischief humor makers. Those that are available and those that can be hacked will be sufficient to generate humorous events for individuals or for groups of city dwellers. In our future, smart environments we will have connected devices and objects that accept traditional input from human users (keyboard, joystick, Wii remote). However, there will be many other ways to provide input, using touch, gestures and information obtained from devices that collect (neuro-) physiological information from a human's body or brain. Smart textiles, using conductive yarn that is woven into the fabric of clothes, and smart materials (Minuto and Nijholt 2013) can also act as sensors and actuators that offer the opportunity to make changes to the appearances and interactive properties of materials, (mobile) objects, social robots, interactive pets, environments, and even human beings who have become smart from a technological point of view and are addressable as well. Townsend (2013) and Cerrudo (2015) focus on Internet and Worldwide Web analogies for smart cities and do not take into account an Internet of Things world, where it is possible to customize, change and control an environment, including devices that happen to be there and characteristics of smart objects and smart humans, including their interaction behaviors. Software agents with physical or virtual embodiments, moving around as social robots, as augmented realities or as holographic humanoids (imaginary people), inhabit our smart environments and can add to their playability but are also vulnerable to hacking, accidental humor and intended but unwanted mischief humor. The same is true for real humans inhabiting smart environments and having sensors and actuators in their pockets or in their clothes, in their smart eyewear, watches or jewelry, on their skin (such as in tattoos), or even in their skulls, brains or other parts of their bodies.

6 Conclusions

In this chapter, we presented our views on the facilitation of humor creation in smart environments. In particular, we looked at humor embedded in games and at attempts to create humorous situations in games. We looked at the behavior of pranksters, cheaters, and hackers and also at legal ways to modify a game in such a way that it becomes less goal-oriented and more oriented toward artistic and mischief-making aims. With smart sensors and actuators, we can design digitally enhanced real worlds that resemble videogame worlds. And as a consequence, we can expect gamers' videogame behavior to also appear in these smart worlds; that is, pranking, trolling, cheating, modifying, and hacking can be expected to occur in smart urban environments, especially given a convergence of game and digitally enhanced real-world environments. This convergence may lead to the introduction of 'game engines' that control parts of our (digitally enhanced) real-world activities. Game-like engines enter and control daily life, in our homes, kitchens and bedrooms, when using public transport and when visiting public places. For this reason, in his enthusiastic talk on gamification, Schell (2010) argues that game designers are needed to design digitally enhanced real worlds.

As argued in this chapter, introducing game elements in the real world will also introduce activities of the smart world or smart city 'gamers.' When they team up, as is happening in multi-player games, they can act as 'smart street' or 'smart city' gangs. 'Flash mobs' that are now organized to take part in the physical world will have their equivalents in the smart world, with participants not only employing social media but also employing a city's digital smartness and playability properties. Such activities will bring humor and fun to the smart city, but we can certainly expect mischievous activities that will be annoying and cause stress and harm to individuals and communities. This will happen despite, or maybe because of, 'smart city protocols' and 'urban operating systems' (Townsend 2013, 289–290).

Acknowledgments I am grateful to Ben Barker (Pan Studio, Fig. 2 Hello Lamp Post), Matt Rosier (Chomko and Rosier, Fig. 3 Shadowing) and Dan Archer (Happy City Lab, Fig. 4, Idea: Dan Archer, Drawings: JP Kalonji) for giving me permission to use pictures from their projects.

References

Berger, A.A.: An Anatomy of Humor. Transaction Publishers, New Brunswick NJ (1993)
Buckels, E.E., Trapnell, P.D., Paulhus, D.L.: Trolls just want to have fun. Pers. Individ. Differ. **67**, 97–102 (2014)

Buijzen, M., Valkenburg, P.: Developing a typology of humor in audiovisual media. Media Psychol.
  6(2), 147–167 (2004)
Callaghan, V., Chin, J., Zamudio, V., Clarke, G., Shahi, A., Gardner M.: Domestic pervasive
  information systems: end-user programming of digital homes Chap. 5. In: Kourouthanassis, P.,
  Giaglis, G. (eds.) Journal of Management Information Systems, Special Edition on Pervasive
  Information Systems, pp. 129–149 (2007)
Carvalho, A., Brisson, A., Paiva, A.: Laugh to me! Implementing emotional escalation on
  autonomous agents for creating a comic sketch. In: Proceedings of the 5th International
  Conference on Interactive Storytelling (ICIDS'12), Lecture Notes in Computer Science, vol.
  7648, pp. 162–173. Springer, Heidelberg (2012). doi:10.1007/978-3-642-34851-8_16
Cavazza, M., Charles, F., Mead, S. J.: Intelligent virtual actors that plan … to fail. In: Smart
  Graphics. Lecture Notes in Computer Science, vol. 2733, pp. 151–161. Springer, Berlin (2003).
  doi:10.1007/3-540-37620-8_15
Cerrudo, C.: *An emerging US (and World) threat: cities wide open to cyber-attacks.* White Paper.
  IOActive, Inc (2015). Retrieved 1 June 2015 from http://www.ioactive.com/pdfs/IOActive_Ha
  ckingCitiesPaper_CesarCerrudo.pdf
Consalvo, M.: Cheating. Gaining Advantage in Videogames. MIT Press, Cambridge, MA (2007)
de Lange, M.: The playful city: using play and games to foster citizen participation. In:
  Skaržauskienė, A. (ed.) Social Technologies and Collective Intelligence, pp. 426–434. Mykolas
  Romeris University, Vilnius (2015)
Dibbell, J.: Mutilated furries, flying phalluses: put the blame on griefers, the sociopaths of the virtual
  world. WIRED Mag. 16(02) (2008). http://archive.wired.com/gaming/virtualworlds/magazine/
  16-02/mf_goons?currentPage=all. Accessed 1 June 2015
Dormann, C.: Fools, tricksters and jokers: categorization of humor in gameplay. In: Reidsma, D.,
  Choi, I., Bargar, R. (eds.) Intelligent Technologies for Interactive Entertainment (INTETAIN
  2014), Lecture Notes of the Institute for Computer Sciences, Social Informatics and
  Telecommunications Engineering, vol. 136, pp. 81–90. Springer, Heidelberg (2014)
Dormann, C., Biddle, R.: A review of humor for computer games: play, laugh and more. Simul.
  Gaming **40**(6), 802–824 (2009)
Freud, S.: Jokes and Their Relation to the Unconscious. W. W. Norton, New York (1905).
  (Republished, 1960)
Grönroos, A.-M.: Humour in video games: play, comedy, and mischief. Master's Thesis, Aalto
  University, School of Art, Design and Architecture, Department of Media, Media Lab, Helsinki
  (2013)
Hulstijn, J., Nijholt, A. (eds.): Computational humor: automatic interpretation and generation of
  verbal humor. In: Proceedings Twente Workshop on Language Technology 12 (TWLT12),
  University of Twente, the Netherlands (1996)
Mihalcea, R.: The multi-disciplinary facets of research on humour. In: Masulli, F., Mitra, S., Pasi
  G. (eds.) WILF 2007, Applications of Fuzzy Sets Theory. Lecture Notes in Artificial
  Intelligence, vol. 4578, pp. 412–421. Springer, Berlin, Heidelberg (2007)
Minsky, M.: Jokes and their relation to the cognitive unconscious. In: Vaina, L., Hintikka, J. (eds.)
  Cognitive Constraints on Communication, pp. 175–200. Reidel, Boston (1981)
Minuto, A., Nijholt, A.: Smart material interfaces as a methodology for interaction. A survey of
  SMIs' state of the art and development. In: 2nd Workshop on Smart Material Interfaces (SMI
  2013). Workshop in Conjunction with 15th ACM International Conference on Multimodal
  Interaction (ICMI'13), ACM, Sidney, NSW, Australia (2013). 13 Dec 2013
Morreal, J.: Taking Laughter Seriously. State University of New York Press, Albany (1983)
Nijholt, A.: Towards humor modelling and facilitation in smart environments. In: Ahram, T. et al.
  (eds.) Proceedings 5th International Conference on Applied Human Factors and Ergonomics
  (AHFE 2014), pp. 2992–3006, Krakow, Poland (2014)
Nijholt, A.: The humor continuum: from text to smart environments. IEEE Proceedings
  International Conference on Informatics, Electronics & Vision (ICIEV), Kitakyushu, Japan,
  pp. 1–10, June, 2015, to appear (2015a)

Nijholt, A.: Designing humor for playable cities. In: Proceedings 6th International Conference on Applied Human Factors and Ergonomics (AHFE 2015), Procedia Manufacturing, vol. 3(C), pp. 2178–2185, Elsevier (ScienceDirect), Las Vegas, USA (2015b)

Olsen, D., Mateas, M.: Beep! Beep! Boom!: towards a planning model of coyote and road runner cartoons. In: Proceedings of the 4th International Conference on Foundations of Digital Games (FDG'09), pp. 145–152. ACM, New York, USA. doi:10.1145/1536513.1536544

Pritchard, M.: How to hurt the hackers: the scoop on internet cheating and how you can combat it. Gamasutra, July 24, 2000. http://www.gamasutra.com/view/feature/131557/how_to_hurt_the _hackers_the_scoop_.php (2000). Accessed 1 June 2015

Raskin, V. (ed.): The Primer of Humor Research. Mouton de Gruyter, Berlin (2008)

Schell, J.: When games invade real life. Talks Best of the Web. Filmed Feb 2010, DICE Summit 2010. http://www.ted.com/talks/jesse_schell_when_games_invade_real_life (2010). Accessed 1 June 2015

Švelch, J.: Comedy of contingency: making physical humor in video game spaces. Int. J. Commun. 8, 2530–2552 (2014)

Thacker, S., Griffiths, M.D.: An exploratory study of trolling in online video gaming. Int J. Cyber Behav. Psychol. Learn (IJCBPL) 2(4), 1–17 (2012). doi:10.4018/ijcbpl.2012100102

Townsend, A.M.: Smart Cities: Big Data, Civic Hackers, and the Quest for a New Utopia, 1st edn. W.W. Norton & Company, New York (2013)

Unger, A.: Modding as part of game culture. In: Fromme, J., Unger, A. (eds.) Computer Games and New Media Cultures. A Handbook of Digital Games Studies, pp. 509–523. Springer, Dordrecht, Netherlands (2012). doi:10.1007/978-94-007-2777-9_32

Warner, D.E., Raiter, M.: Social context in massively-multiplayer online games (MMOGs): ethical questions in shared space. Int. Rev. Inf. Ethics 4(12), 46–52 (2005)

Printed in the United States
By Bookmasters